SPELLMAN'S STANDARD HANDBOOK FOR WASTEWATER OPERATORS

HOW TO ORDER THIS BOOK

BY PHONE: 800-233-9936 or 717-291-5609, 8AM–5PM Eastern Time

BY FAX: 717-295-4538

BY MAIL: Order Department
Technomic Publishing Company, Inc.
851 New Holland Avenue, Box 3535
Lancaster, PA 17604, U.S.A.

BY CREDIT CARD: American Express, VISA, MasterCard

BY WWW SITE: http://www.techpub.com

PERMISSION TO PHOTOCOPY—POLICY STATEMENT

Authorization to photocopy items for internal or personal use, or the internal or personal use of specific clients, is granted by Technomic Publishing Co., Inc. provided that the base fee of US $3.00 per copy, plus US $.25 per page is paid directly to Copyright Clearance Center, 222 Rosewood Drive, Danvers, MA 01923, USA. For those organizations that have been granted a photocopy license by CCC, a separate system of payment has been arranged. The fee code for users of the Transactional Reporting Service is 1-56676/99 $5.00 + $.25.

VOLUME 1
FUNDAMENTAL LEVEL

Spellman's Standard Handbook for Wastewater Operators

Frank R. Spellman, Ph.D.

TECHNOMIC PUBLISHING CO., INC.
LANCASTER · BASEL

Fundamental Level, Volume 1
a TECHNOMIC publication

Technomic Publishing Company, Inc.
851 New Holland Avenue, Box 3535
Lancaster, Pennsylvania 17604 U.S.A.

Copyright © 1999 by Technomic Publishing Company, Inc.
All rights reserved

No part of this publication may be reproduced, stored in a
retrieval system, or transmitted, in any form or by any means,
electronic, mechanical, photocopying, recording, or otherwise,
without the prior written permission of the publisher.

Printed in the United States of America
10 9 8 7 6 5 4 3 2 1

Main entry under title:
    Spellman's Standard Handbook for Wastewater Operators—Fundamental Level, Volume 1

A Technomic Publishing Company book
Bibliography: p.
Includes index p. 271

Library of Congress Catalog Card No. 99-61165
ISBN No. 1-56676-741-5

For Wastewater Operators Everywhere

Table of Contents

Preface xi
Acknowledgements xiii

1. INTRODUCTION 1

    1.1   Introduction   1
    1.2   The Wastewater Treatment Process: The Model   2
    1.3   The Scope of Text   3

2. WASTEWATER TERMINOLOGY AND DEFINITIONS 5

    2.1   Introduction   5
    2.2   Terminology and Definitions   5
    2.3   Reference   8

3. BASIC WASTEWATER MATHEMATICS 9

    3.1   Introduction   9
    3.2   Calculators   10
    3.3   Basic Math   11
    3.4   Units of Measurement   33
    3.5   Circumference, Areas, and Volumes   34
    3.6   Geometric Mean   42
    3.7   Moving Average   43
    3.8   Mass Balance   44
    3.9   References   47
    3.10 Chapter Review Questions   47

4. CONVERSIONS 51

    4.1   Conversion Factors   51
    4.2   The Conversion Table   53
    4.3   Chapter Review Questions   59

5. MEASURING PLANT PERFORMANCE 61

    5.1   Introduction   61
    5.2   Plant Performance/Efficiency   61
    5.3   Unit Process Performance/Efficiency   61
    5.4   Percent Volatile Matter Reduction in Sludge   62
    5.5   Chapter Review Questions   62

6. HYDRAULIC DETENTION TIME 65

    6.1   Introduction   65

    6.2   Detention Time in Days   65
    6.3   Detention Time in Hours   65
    6.4   Detention Time in Minutes   66
    6.5   Chapter Review Questions   66

7. WASTEWATER: SOURCES AND CHARACTERISTICS 69

    7.1   Introduction   69
    7.2   Wastewater Sources   69
    7.3   Wastewater Characteristics   70
    7.4   Typical Domestic Wastewater Characteristics   72
    7.5   Chapter Review Questions   73

8. WASTEWATER BIOLOGY 75

    8.1   Introduction   75
    8.2   Wastewater Organisms   75
    8.3   Biological Processes   77
    8.4   The Growth Curve   79
    8.5   Self- (Natural) Purification   80
    8.6   Biogeochemical Cycles   81
    8.7   Requirements for Biological Activity   82
    8.8   References   84
    8.9   Chapter Review Questions   84

9. WATER HYDRAULICS 85

    9.1   Introduction   85
    9.2   Basic Concepts   85
    9.3   Head   89
    9.4   Flow   91
    9.5   Reference   91
    9.6   Chapter Review Questions   92

10. PUMPS 93

    10.1   Introduction   93
    10.2   Types of Pumps   93
    10.3   Characteristics: Centrifugal and Positive Displacement Pumps   94
    10.4   Operating Pumps   94
    10.5   Pump Calculations   96
    10.6   References   99
    10.7   Chapter Review Questions   99

11. WASTEWATER COLLECTION SYSTEMS 101

    11.1   Wastewater Collection Systems   101
    11.2   Pumping Stations   103
    11.3   Pumping Station Wet Well Calculations   105
    11.4   Chapter Review Question   107

12. PRELIMINARY TREATMENT 109

    12.1   Introduction   109
    12.2   Screening   109
    12.3   Shredding   112
    12.4   Grit Removal   113

12.5 Flow Measurement   116
12.6 Preaeration   118
12.7 Chemical Addition   119
12.8 Flow Equalization   119
12.9 Reference   119
12.10 Chapter Review Questions   119

13. SEDIMENTATION 121

13.1 Introduction   121
13.2 Process Description   121
13.3 Types of Sedimentation Tanks   121
13.4 Process Control Calculations   125
13.5 Effluent from Settling Tanks   127
13.6 Reference   127
13.7 Chapter Review Questions   127

14. SECONDARY TREATMENT: PONDS, TRICKLING FILTERS, AND RBCs 131

14.1 Introduction   131
14.2 Secondary Treatment   131
14.3 Treatment Ponds   132
14.4 Trickling Filters   137
14.5 Rotating Biological Contactors   143
14.6 Chapter Review Questions   146

15. ACTIVATED SLUDGE 151

15.1 Introduction   151
15.2 Activated Sludge Process: Operation of   151
15.3 Activated Sludge Process: Equipment   152
15.4 Activated Sludge Process: Modifications   153
15.5 Aeration Tank Observations   154
15.6 Final Settling Tank (Clarifier) Observations   155
15.7 Process Control Testing and Sampling   156
15.8 Process Controls   158
15.9 Troubleshooting Operational Problems   159
15.10 Process Control Calculations   161
15.11 Solids Concentration: Secondary Clarifier   166
15.12 Activated Sludge Process Recordkeeping Requirements   166
15.13 Reference   166
15.14 Chapter Review Questions   166

16. DISINFECTION: CHLORINATION/DECHLORINATION 169

16.1 Introduction   169
16.2 Chlorination   169
16.3 Process Calculations   174
16.4 Troubleshooting Operational Problems   178
16.5 Dechlorination   181
16.6 Reference   181
16.7 Chapter Review Questions   181

17. PROCESS RESIDUALS (SLUDGE) TREATMENT 183

17.1 Introduction   183
17.2 Sludge Pumping Calculations   183

17.3 Sludge Thickening   187
17.4 Stabilization   189
17.5 Sludge Dewatering   199
17.6 Chapter Review Questions   202

18. WASTEWATER SAMPLING AND TESTING 205

18.1 Introduction   205
18.2 Wastewater Sampling   205
18.3 Wastewater Testing Methods   208
18.4 Chapter Review Questions   226

19. PERMITS, RECORDS, AND REPORTS 229

19.1 Introduction   229
19.2 Definitions   229
19.3 NPDES Permits   230
19.4 Chapter Review Questions   234

20. FINAL REVIEW EXAM 237

20.1 Introduction   237
20.2 Review Exam   237

Appendix A: Answers to Chapter Review Questions   253
Appendix B: Answers to Final Review Exam: Chapter 20   263
Appendix C: Commonly Used Formulae in Wastewater Treatment   269
Index   271

Preface

SPELLMAN'S *Standard Handbook for Wastewater Operators* is more than just a three-volume study guide or a readily accessible source of information for review in preparing wastewater personnel for operator certification and licensure. This three-volume handbook is a resource manual and a troubleshooting guide that contains a compilation of wastewater treatment information, data, operational material, process control procedures and problem solving, safety and health information, new trends in wastewater treatment administration and technology, and numerous sample problem-solving practice sets. The text's most important aspects are threefold:

(1) It gives today's wastewater operators instant information they need to expand their knowledge—which aids them in the efficient operation of a wastewater treatment plant.
(2) It provides the user with the basic information and sample problem-solving sets needed to prepare for state licensing/certification examinations.
(3) It provides user-friendly, straightforward, plain English fundamental reference material—a three-volume handbook of information and unit process troubleshooting guidance required on a daily basis, not only by the plant manager, plant superintendent, chief operator, lab technician, and maintenance operator, but more importantly by and for the plant operator.

We could say that the handbook's primary goal is to enhance the understanding, awareness, and abilities of practicing operators and those who aspire to be operators. The first volume (introductory), second volume (intermediate), and third volume (advanced) are designed to build on one another, providing increasingly advanced information.

The message conveyed by this handbook is quite simple: None of us are chained to the knowledge we already have—we should strive to increase our technical knowledge and expertise constantly. For those preparing for operator's licensing, this is critical; wastewater treatment is a complex process. For those seasoned, licensed, veteran operators, continuous review is also critical, because wastewater treatment is still an evolving, dynamic, ever-changing field. Spellman's handbook series (which we think of as "Answer Books") provides the vehicle for reaching these goals.

In short, preparing for qualification as a wastewater treatment plant operator and providing a quick, ready reference for those who have already obtained their licenses are what this handbook is all about.

Is this text needed? Absolutely.

Contrary to popular belief (and simply put), treating wastewater is not just an art, but instead is both an art and a science. Treating wastewater successfully demands technical expertise and a broad range of available technologies, as well as an appreciation for and the understanding of the fundamental environmental and health reasons for the processes involved. It demands unique vision and capabilities. This is where *Spellman's Standard Handbook for Wastewater Operators* comes in. From pumping influent, treating the wastestream, through managing biosolids, this handbook series provides easy to understand state-of-the-art information in a three-volume set that begins at the fundamental level for those preparing for the Class IV/III or Grade I/II operator examination, proceeds to the intermediate level for the Class III/II or Grade II/III operator examination, and finishes at the advanced level for Class I/Grade IV/V wastewater operator license examination. Though formatted at three separate levels (introductory, intermediate, and advanced), overlap between each volume ensures continuity and a smooth read from volume to volume. In essence, each volume is a handheld reference text—one that enables the practitioner of the artful science of wastewater treatment to qualify for certification and/or refresh his or her memory in an easy, precise, efficient, effective manner.

This handbook was prepared to help operators obtain licensing and operate wastewater treatment plants properly. It can also be used as a textbook in technical training courses in technical schools and at the junior college level.

Note that the handbook does not discuss the specific content of the examination. It reviews the wastewater operator's job-related knowledge identified by the examination developers as essential for a minimally competent Class IV through Class I or Grade I through Grade V wastewater treatment plant operator. Every attempt has been made to make the three-volume handbook as comprehensive as possible while maintaining its compact, practical format.

The bottom line: The handbook is not designed to simply "teach" the operator licensing exams, although it is immediately obvious to the users that the material presented will help them pass licensing exams. The material in each volume is intended for practical use and application. We present applied math and chemistry by way of real-world problems. Readers learn how to maintain equipment. We explain apparatus used in the laboratory and in the field (e.g., valves and pumps).

Will the handbook series help you obtain a passing score on certification exams? Yes. If you follow it, use it, and reuse it, it will help—and this is the real bottom line.

Acknowledgements

ALTHOUGH this three-volume set of handbooks bears my name, it should be pointed out that it is a compilation of many individuals' efforts put forth over the years. Modeled after the highly successful Wastewater Treatment Plant Operators' Short Course presented annually by the Virginia Department of Environmental Quality at Virginia Polytechnic Institute and State University (Virginia Tech), this handbook series has benefited from many contributors, including students—too numerous in number to acknowledge here.

When one finds a model, a prototype, a paradigm—one that actually works—one that actually enables students to gain from the learning experience and to go on to have successful careers in the wastewater field—it seems only natural and quite fitting to bottle up such a program and make it available to wastewater practitioners worldwide. This is the purpose of this handbook series—to provide a model that works; we know this because it has been successfully tested, time and time again.

The format and material contained within this three-volume set have been continuously updated and used successfully for more than 20 years to train wastewater personnel for licensure and to equip them with the requisite knowledge required to operate wastewater treatment plants in the most efficient manner possible.

In short, though there are too many individuals to single out for recognition for their contributions to this text, there are always a few who stand out above the rest—and this is certainly the case with this publication. Thus, Dr. Gregory D. Boardman, an associate professor of environmental engineering at Virginia Tech and Wayne Staples of Virginia's Department of Environmental Quality (DEQ) deserve much credit for the format and material contained within each of the three handbooks. Without their efforts and expertise, this handbook series would not have been possible.

CHAPTER 1

Introduction

What is unsought will go undetected.—Sophocles

1.1 INTRODUCTION

SPELLMAN'S Standard Handbook for Wastewater Operators, Volume 1: Fundamental Level is primarily designed to provide a readily accessible, user-friendly source of information for review in preparing for the Class IV/III or Grade I/II state Wastewater Operator Licensure Examinations. Along with providing the necessary information to help the user successfully pass the Class IV/III or Grade I/II examinations, Volume 1 sets the stage (provides the basics) for both Volumes 2 and 3, which are intended to prepare users to sit for examinations for Class II/I or Grade III/IV/V licensure.

Every attempt has been made to format this presentation in a way that allows the user to build upon information presented, step-by-step, page-by-page, as he or she progresses through the material. This handbook represents a summary of expert information available in many other sources (see Table 1.1). For additional information or more specific material on any of the topics presented, the user is advised to consult one or more of the references provided in Table 1.1.

This fundamental level handbook assumes that the user is an operator-in-training who is currently preparing to sit for the Class IV/III or Grade I/II operator licensure examination.

✓ *Note:* In this handbook, we refer to the "fundamental level" as those first two steps in licensure, which is the case in many states. [The symbol ✓ ("check mark") displayed in various locations throughout the handbook indicates an important point or note that the reader should read carefully.]

Those people with limited experience who do not qualify to sit for subject examinations may find the material helpful but should augment the content of this handbook with other, more in-depth training such as the various field study programs available from state water control boards, short courses presented by various universities (e.g., Virginia Tech) and/or technical schools, and correspondence studies from such sources as California State University, Sacramento (The "Sacramento Manuals").

It is important to point out that changes in technology and regulations occur frequently in the water pollution control industry. Because of this, it is important for the licensure candidate to stay abreast of these changes.

The handbook is divided into chapters by sections covering specific topic areas. At the end of many chapters, a series of review questions is included. Upon completion of these chapters, answer the review questions, and check your answers with those given in Appendix A. The final chapter of the handbook includes a comprehensive practice examination. The purpose of the comprehensive practice examination is to test the level of knowledge the user has attained through study of this handbook, knowledge gained through on-the-job experience, and knowledge gained from other

TABLE 1.1. Recommended Reference Material.

1. *Advanced Waste Treatment, A Field Study Program,* 2nd ed., Kerri, K., et al. California State University, Sacramento, CA.
2. *Aerobic Biological Wastewater Treatment Facilities*, Environmental Protection Agency, EPA 430/9-77-006, Washington, D.C., 1977.
3. *Anaerobic Sludge Digestion*, Environmental Protection Agency, EPA 430/9-76-001, Washington, D.C., 1977.
4. *Annual Book of ASTM Standards, Section 11, "Water and Environmental Technology,"* American Society for Testing Materials (ASTM), Philadelphia, PA.
5. *Guidelines Establishing Test Procedures for the Analysis of Pollutants.* Federal Register (40 CFR 136), April 4, 1995, Volume 60, No. 64, Page 17160.
6. *Handbook of Water Analysis*, 2nd ed., HACH Chemical Company, P.O. Box 389, Loveland, CO, 1992.
7. *Industrial Waste Treatment, A Field Study Program, Volume I*, Kerri, K., et al. California State University, Sacramento, CA.
8. *Industrial Waste Treatment, A Field Study Program, Volume 2,* Kerri, K., et al. California State University, Sacramento, CA.
9. *Methods for Chemical Analysis of Water and Wastes,* U.S. Environmental Protection Agency, Environmental Monitoring Systems Laboratory-Cincinnati (EMSL-CL), EPA-6000/4-79-020, Revised March 1983 and 1979 (where applicable).
10. *O & M of Trickling Filters, RBC and Related Processes, Manual of Practice OM-10*, Water Pollution Control Federation (now called Water Environment Federation), Alexandria, VA, 1988.
11. *Operation of Wastewater Treatment Plants, A Field Study Program, Volume I*, 4th ed., Kerri, K., et al. California State University, Sacramento, CA.
12. *Operation of Wastewater Treatment Plants, A Field Study Program, Volume II*, 4th ed., Kerri, K., et al. California State University, Sacramento, CA.
13. *Standard Methods for the Examination of Water and Wastewater*, 18th ed., American Public Health Association, American Water Works Association-Water Environment Federation, Washington, D.C., 1992.
14. *Treatment of Metal Wastestreams*, Kerri, K., et al. California State University, Sacramento, CA.
15. *Basic Math Concepts: For Water and Wastewater Plant Operators.* Price, J. K. Lancaster, PA: Technomic Publishing Company, Inc., 1991.
16. *Simplified Wastewater Treatment Plant Operations.* Haller, E. J. Lancaster, PA: Technomic Publishing Company, Inc., 1995.
17. *Wastewater Treatment Plants: Planning, Design, and Operation,* 2nd ed. Qasim, S. R. Lancaster, PA: Technomic Publishing Company, Inc., 1999.

sources. A score of 75% or above is considered "good"—more importantly, any questions missed signal to the user the need to go back and reread and re-study the applicable areas. By using the final examination as a measuring stick, the user can gauge his or her level of knowledge in all pertinent areas and determine strong and weak points. When you get right down to it, shouldn't this be the purpose of any examination? That is, to measure one's level of knowledge in such a way so as to point to the proper direction to take to attain an even greater level of knowledge. This seems like a worthwhile objective, and it is. Answers to the final review examination in Chapter 20 are provided in Appendix B. A formula sheet is provided in Appendix C and should be used for reference; it can and should be used when taking the final examination.

1.2 THE WASTEWATER TREATMENT PROCESS: THE MODEL

Figure 1.1 shows a basic schematic of a sample wastewater treatment process providing primary and secondary treatment using the activated sludge process. This is the model, the prototype, the

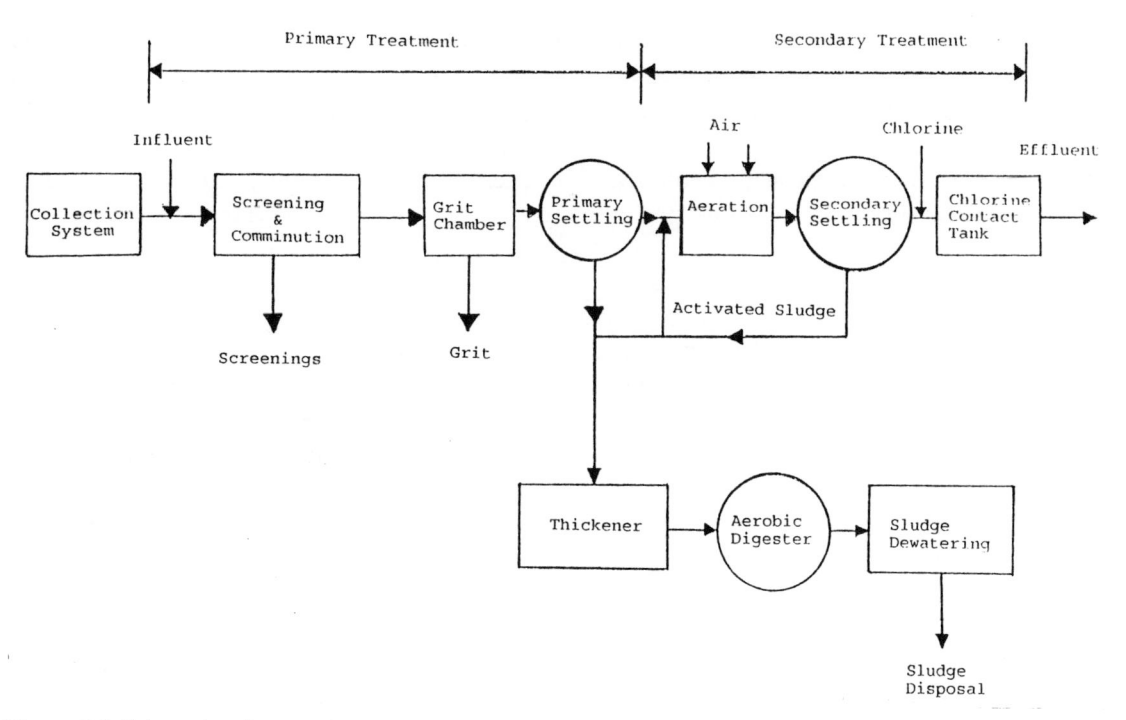

Figure 1.1 Schematic of a sample wastewater treatment process providing primary and secondary treatment using activated sludge process.

paradigm used in all three volumes of this handbook. Though it is true that in secondary treatment (which provides biochemical oxygen demand removal beyond what is achievable by simple sedimentation) there are actually three commonly used approaches (trickling filter, activated sludge, and oxidation ponds), we mainly focus on the activated sludge process throughout this handbook primarily for instructive and illustrative purposes. The purpose of Figure 1.1 and its subsequent renditions is to allow the reader to follow the treatment process step-by-step as it is presented (and as it is actually configured in the real world) in the written material—to aid in gaining understanding of how all the various unit processes sequentially follow and tie into each other. Therefore, in Volume 1, we begin certain chapters (which discuss unit processes) with Figure 1.1—with the relevant subject area to be discussed included in the diagram along with previously presented processes. In essence, what we are doing is starting with a blank diagram and filling in the unit processes as we progress. It is important to begin these chapters in this manner because wastewater treatment is a series of individual steps (unit processes) that treat the wastestream as it makes its way through the entire process—thus, it logically follows that a pictorial presentation along with pertinent written information should enhance the learning process. It should also be pointed out, however, that even though the model shown in Figure 1.1 does not include all unit processes currently used in wastewater treatment, we do not ignore the other major processes: trickling filters, rotating biological contactors (RBCs), and oxidation ponds. These are given their due and are presented and discussed in some detail.

1.3 THE SCOPE OF TEXT

Volume 1 of the handbook series consists of 20 chapters. Along with the introduction presented in Chapter 1, Chapter 2 explains the pertinent terminology and definitions used in wastewater treatment. Chapter 3 is a comprehensive discussion of wastewater mathematics. The wastewater operator soon learns that he or she can't do his or her job correctly, effectively, and precisely unless he or she is well-versed in basic mathematics.

✓ Note that mastering the basic mathematical principles included in Volume 1 plays an important role in process control calculations presented in advanced materials in Handbook Volumes 2 and 3.

Chapter 4 explains how to use conversion factors. This is important because, along with a working knowledge of basic math, the wastewater worker must be able to use conversion factors to change measurements or calculated values from one unit of measure to another. Chapter 5 explains how to measure (mathematically) plant performance/efficiency (or percent removal). The percent removal calculation is used throughout the treatment process to evaluate how well a plant or unit treatment process is operating—obviously, these are critical parameters. Chapter 6 addresses Hydraulic Detention Time (HDT)—the theoretical time the wastewater remains in a treatment unit. Chapter 7 discusses wastewater sources and specific characteristics. Chapter 8 presents a brief overview of wastewater biology. Wastewater contains a large variety of microorganisms; thus, a fundamental knowledge of wastewater biology (microbiology) is important to the operator. Chapter 9 discusses the study of how liquids act as they move through a channel or a pipe. Chapter 10 explains the movement of wastewater from the source, through the collection and treatment system to the discharge point, which requires energy. This energy is supplied by pumps; thus, a basic knowledge of pumps and their conduits and piping is important. Chapter 11 explains the function of wastewater collection systems. Preliminary treatment is presented in Chapter 12. Chapter 13 discusses the sedimentation process. Chapter 14 explains the secondary treatment processes, including treatment ponds, trickling filters, and RBCs. The activated sludge process is discussed in Chapter 15. Chapter 16 explains disinfection. Wastewater process residuals (sludge) are discussed in Chapter 17. The monitoring and controlling of wastewater processes, provided by sampling and testing (evaluation), is discussed in Chapter 18. Permits, records, and reports and their importance are discussed in Chapter 19. Chapter 20, the final chapter, contains the final review examination.

From the information provided to this point, it should be obvious to the reader that the material in Volume 1 of the handbook is presented in a logical, step-by-step manner—which not only works to aid those who are studying to sit for licensure, but is also helpful to those who use this handbook as a ready reference or as a troubleshooting guide; that is, as an answer book. Speaking of answers—in answer to Sophocles' statement, which opened this chapter, our counter response is, this handbook series makes information (answers) available; they only require detection. Detection is, of course, accomplished through use.

CHAPTER 2

Wastewater Terminology and Definitions

If you wish to converse with me, define your terms.—Voltaire

2.1 INTRODUCTION

EVERY branch of science and technology has its own terms with their own accompanying definitions. Many of the terms used are unique, others combine words from many different professions. Why is this the case? Can't we just simplify things and use the same words with the same meaning? Yes, we can do this—and we often do. For example, when we have a conversation in English with someone who understands and speaks English, we are using the same words that, hopefully, both parties understand to have basically the same meaning. The problem occurs when two or more individuals having a conversation in English are from different professions. For example, let's assume that one of these individuals is a medical doctor, one is a lawyer, and the other is an anthropologist. In general conversation, none of these three would have a problem understanding each other. However, if the medical doctor starts to talk medicine, the lawyer begins to speak legalese, and the anthropologist talks about anthropology, unless each is familiar with the other's terms (and their meaning), they would have difficulty communicating (on the technical level) with each other.

Wastewater treatment technology, like many other technical fields, has its own unique terms with their own meanings. Though some of the terms are unique, many are common to other professions. Remember, the science of wastewater treatment is a combination of engineering, biology, mathematics, hydrology, chemistry, physics, and other disciplines. Therefore, many of the terms used in engineering, biology, mathematics, hydrology, chemistry, physics, and others are also used in wastewater treatment.

To enhance the learning activity and to facilitate speaking in the language of wastewater treatment, this chapter identifies and defines many of the terms unique to wastewater treatment. Those terms not listed or defined in the following section will be defined as they appear in the text.

2.2 TERMINOLOGY AND DEFINITIONS

- *Activated sludge* the solids formed when microorganisms are used to treat wastewater using the activated sludge treatment process. It includes organisms, accumulated food materials, and waste products from the aerobic decomposition process.
- *Advanced wastewater treatment* treatment technology to produce an extremely high-quality discharge.
- *Aerobic* conditions in which free, elemental oxygen is present. Also used to describe organisms, biological activity, or treatment processes that require free oxygen.
- *Anaerobic* conditions in which no oxygen (free or combined) is available. Also used to

describe organisms, biological activity, or treatment processes that function in the absence of oxygen.
- *Anoxic* conditions in which no free, elemental oxygen is present. The only source of oxygen is combined oxygen, such as that found in nitrate compounds. Also used to describe biological activity or treatment processes that function only in the presence of combined oxygen.
- *Average monthly discharge limitation* the highest allowable discharge over a calendar month.
- *Average weekly discharge limitation* the highest allowable discharge over a calendar week.
- *Biochemical oxygen demand, BOD_5* the amount of organic matter that can be biologically oxidized under controlled conditions (5 days @ 20°C in the dark).
- *Biosolids* from *Merriam-Webster's Collegiate Dictionary, Tenth Edition* (1998): biosolid *n* (1977)—solid organic matter recovered from a sewage treatment process and used especially as fertilizer—usually used in plural. *Note:* In this text and all other Spellman texts on wastewater topics, biosolids is used in many places (activated sludge being the exception) to replace the standard term sludge. The author (along with others in the field) views the term sludge as an ugly four-letter word that is inappropriate to use in describing biosolids. Biosolids are a product that can be reused; they have some value. Because biosolids have some value, they certainly should not be classified as a "waste" product—and in subsequent volumes (Volumes 2 and 3) of this handbook, when biosolids for beneficial reuse is addressed, it is made clear that biosolids is not a waste product.
- *Buffer* a substance or solution that resists changes in pH.
- *Carbonaceous biochemical oxygen demand, $CBOD_5$* the amount of biochemical oxygen demand that can be attributed to carbonaceous material.
- *Chemical oxygen demand (COD)* the amount of chemically oxidizable materials present in the wastewater.
- *Clarifier* a device designed to permit solids to settle or rise and be separated from the flow. Also known as a settling tank or sedimentation basin.
- *Coliform* a type of bacteria used to indicate possible human or animal contamination of water.
- *Combined sewer* a collection system that carries both wastewater and stormwater flows.
- *Comminution* a process to shred solids into smaller, less harmful particles.
- *Composite sample* a combination of individual samples taken in proportion to flow.
- *Daily discharge* the discharge of a pollutant measured during a calendar day or any 24-hour period that reasonably represents a calendar day for the purposes of sampling. Limitations expressed as weight are total mass (weight) discharged over the day. Limitations expressed in other units are average measurements of the day.
- *Daily maximum discharge* the highest allowable values for a daily discharge.
- *Detention time* the theoretical time water remains in a tank at a given flow rate.
- *Dewatering* the removal or separation of a portion of water present in a sludge or slurry.
- *Discharge monitoring report (DMR)* the monthly report required by the treatment plant's National Pollutant Discharge Elimination System (NPDES) discharge permit.
- *Dissolved oxygen (DO)* free or elemental oxygen that is dissolved in water.
- *Effluent* the flow leaving a tank, channel, or treatment process.
- *Effluent limitation* any restriction imposed by the regulatory agency on quantities, discharge rates, or concentrations of pollutants that are discharged from point sources into state waters.
- *Facultative* organisms that can survive and function in the presence or absence of free, elemental oxygen.

- *Fecal coliform*   a type of bacteria found in the bodily discharges of warm-blooded animals. Used as an indicator organism.
- *Floc*   solids that join together to form larger particles that will settle better.
- *Flume*   a flow rate measurement device.
- *Food-to-microorganism ratio (F/M)*   an activated sludge process control calculation based upon the amount of food (BOD$_5$ or COD) available per pound of mixed liquor volatile suspended solids.
- *Grab sample*   an individual sample collected at a randomly selected time.
- *Grit*   heavy inorganic solids, such as sand, gravel, egg shells, or metal filings.
- *Industrial wastewater*   wastes associated with industrial manufacturing processes.
- *Infiltration/inflow*   extraneous flows in sewers; defined by Metcalf & Eddy (1991), pp. 29–31 as follows:
  —*Infiltration*   water entering the collection system through cracks, joints, or breaks.
  —*Steady inflow*   water discharged from cellar and foundation drains, cooling water discharges, and drains from springs and swampy areas. This type of inflow is steady and is identified and measured along with infiltration.
  —*Direct flow*   those types of inflow that have a direct stormwater runoff connection to the sanitary sewer and cause an almost immediate increase in wastewater flows. Possible sources are roof leaders, yard and areaway drains, manhole covers, cross connections from storm drains and catch basins, and combined sewers.
  —*Total inflow*   the sum of the direct inflow at any point in the system plus any flow discharged from the system upstream through overflows, pumping station bypasses, and the like.
  —*Delayed inflow*   stormwater that may require several days or more to drain through the sewer system. This category can include the discharge of sump pumps from cellar drainage as well as the slowed entry of surface water through manholes in ponded areas.
- *Influent*   the wastewater entering a tank, channel, or treatment process.
- *Inorganic*   mineral materials, such as salt, ferric chloride, iron, sand, gravel, etc.
- *License*   a certificate issued by the State Board of Waterworks/Wastewater Works Operators authorizing the holder to perform the duties of a wastewater treatment plant operator.
- *Mean cell residence time (MCRT)*   the average length of time a mixed liquor suspended solids particle remains in the activated sludge process. May also be known as sludge retention time.
- *Mixed liquor*   the combination of return activated sludge and wastewater in the aeration tank.
- *Mixed liquor suspended solids (MLSS)*   the suspended solids concentration of the mixed liquor.
- *Mixed liquor volatile suspended solids (MLVSS)*   the concentration of organic matter in the mixed liquor suspended solids.
- *Milligrams/liter (mg/L)*   a measure of concentration. It is equivalent to parts per million (ppm).
- *Nitrogenous oxygen demand (NOD)*   a measure of the amount of oxygen required to biologically oxidize nitrogen compounds under specified conditions of time and temperature.
- *NPDES permit*   National Pollutant Discharge Elimination System permit, which authorizes the discharge of treated wastes and specifies the condition that must be met for discharge.
- *Nutrients*   substances required to support living organisms. Usually refers to nitrogen, phosphorus, iron, and other trace metals.
- *Organic*   materials that consist of carbon, hydrogen, oxygen, sulfur, and nitrogen. Many

organics are biologically degradable. All organic compounds can be converted to carbon dioxide and water when subjected to high temperatures.
- *Pathogenic* — disease causing. A pathogenic organism is capable of causing illness.
- *Point source* — any discernible, defined, and discrete conveyance from which pollutants are or may be discharged.
- *Part per million (ppm)* — an alternative (but numerically equivalent) unit used in chemistry is milligrams per liter (mg/L). As an analogy, think of a ppm as being equivalent to a full shot glass in a swimming pool.
- *Return activated sludge solids (RASS)* — the concentration of suspended solids in the sludge flow being returned from the settling tank to the head of the aeration tank.
- *Sanitary wastewater* — wastes discharged from residences and from commercial, institutional, and similar facilities that include both sewage and industrial wastes.
- *Scum* — the mixture of floatable solids and water that is removed from the surface of the settling tank.
- *Septic* — a wastewater that has no dissolved oxygen present. Generally characterized by black color and rotten egg (hydrogen sulfide) odors.
- *Settleability* — a process control test used to evaluate the settling characteristics of the activated sludge. Readings taken at 30 to 60 minutes are used to calculate the settled sludge volume (SSV) and the sludge volume index (SVI).
- *Settled sludge volume* — the volume in percent occupied by an activated sludge sample after 30 to 60 minutes of settling. Normally written as SSV with a subscript to indicate the time of the reading used for calculation (SSV_{60} or SSV_{30}).
- *Sewage* — wastewater containing human wastes.
- *Sludge* — the mixture of settleable solids and water that is removed from the bottom of the settling tank.
- *Sludge retention time (SRT)* — *see* mean cell residence time.
- *Sludge volume index (SVI)* — a process control calculation that is used to evaluate the settling quality of the activated sludge. Requires the SSV_{30} and mixed liquor suspended solids test results to calculate.
- *Storm sewer* — a collection system designed to carry only stormwater runoff.
- *Stormwater* — runoff resulting from rainfall and snowmelt.
- *Supernatant* — in a digester, it is the amber-colored liquid above the sludge.
- *Wastewater* — the water supply of the community after it has been soiled by use.
- *Waste activated sludge solids (WASS)* — the concentration of suspended solids in the sludge, which is being removed from the activated sludge process.
- *Weir* — a device used to measure wastewater flow.
- *Zoogleal slime* — the biological slime that forms on fixed film treatment devices. It contains a wide variety of organisms essential to the treatment process.

2.3 REFERENCE

Metcalf & Eddy. *Wastewater Engineering: Treatment, Disposal, Reuse*, 3rd. ed., New York: McGraw-Hill, Inc., 1991.

CHAPTER 3

Basic Wastewater Mathematics

When studying a discipline that does not include mathematics, one thing is certain: the discipline under study has nothing or little to do with science. (Spellman, The Science of Air, *1999)*

3.1 INTRODUCTION

ANYONE who has had the opportunity to work in wastewater treatment, for even a short time, quickly learns that wastewater treatment operations involve a large number of process control calculations. All of these calculations are based upon basic math principles. In this chapter, we cover basic mathematical operations that wastewater operators are required to use, many of them on a daily basis. Along with an explanation of each math operation, several sample problems are included. A practice test is included at the end of this chapter to determine your level of math skill—this is important because, without mastering basic math skills, the user of this text will find subsequent chapters (which include a lot of math) difficult to complete.

For those with a fear of math (and apparently there are a large number in this category), let's take a look at how one seasoned math teacher puts it: "Those who have difficulty in math often do not lack the ability for mathematical calculation, they merely have not learned, or have been taught, the 'language of math' " (Price, 1991, p. vii).

Price's point (the "language of math") is well taken, and it can be expanded somewhat to "The language of mathematics is a universal language." Mathematical symbols have the same meaning to people speaking in many different languages throughout the world.

What is mathematics? Good question. Mathematics is numbers. Math uses combinations of numbers and symbols to solve practical problems. Every day, you use numbers to count. Numbers may be considered as representing things counted. The money in your pocket or the power consumed by an electric motor are expressed in numbers. Again, we use numbers every day. Because we use numbers every day, we are all mathematicians, to a point.

In wastewater treatment, we need to take math beyond "to a point." We need to learn, understand, appreciate, and use mathematics. Probably the greatest single cause of failure to understand and appreciate mathematics is not knowing the key definitions of the terms used. In mathematics, more than in any other subject, each word used has a definite and fixed meaning.

The following basic definitions will aid you in understanding the material that follows.

3.1.1 MATH TERMINOLOGY AND DEFINITIONS

An *integer*, or an *integral number*, is a whole number. Thus 1, 2, 3, 4, 5, 6, 7, 8, 9, 10, 11, and 12 are the first 12 positive integers.

A *factor*, or *divisor*, of a whole number is any other whole number that exactly divides it. Thus, 2 and 5 are factors of 10.

A *prime number* in math is a number that has no factors except itself and 1. Examples of prime numbers are 1, 3, 5, 7, and 11.

A *composite number* is a number that has factors other than itself and 1. Examples of composite numbers are 4, 6, 8, 9, and 12.

A *common factor*, or *common divisor*, of two or more numbers is a factor that will exactly divide each of them. If this factor is the largest factor possible, it is called the *greatest common divisor*. Thus, 3 is a common divisor of 9 and 27, but 9 is the greatest common divisor of 9 and 27.

A *multiple* of a given number is a number that is exactly divisible by the given number. If a number is exactly divisible by two or more other numbers, it is a common multiple of them. The least (smallest) such number is called the *lowest common multiple*. Thus, 36 and 72 are common multiples of 12, 9, and 4; however, 36 is the lowest common multiple.

An *even number* is a number exactly divisible by 2. Thus, 2, 4, 6, 8, 10, and 12 are even integers.

An *odd number* is an integer that is not exactly divisible by 2. Thus, 1, 3, 5, 7, 9, and 11 are odd integers.

A *product* is the result of multiplying two or more numbers together. Thus, 25 is the product of 5×5. Also, 4 and 5 are factors of 20.

A *quotient* is the result of dividing one number by another. For example, 5 is the quotient of 20 divided by 4.

A *dividend* is a number to be divided; a *divisor* is a number that divides. For example, in $100 \div 20 = 5$, 100 is the dividend, 20 is the divisor, and 5 is the quotient.

3.2 CALCULATORS

There can be little doubt that the proper use of a calculator can reduce time and effort required to perform calculations. Thus, it is important to recognize the calculator as a helpful tool—but not the total answer. First and most important in selecting a calculator is to ensure that it has a well-illustrated instruction manual. The manual should be large enough to read, not an inch by an inch and a quarter. It should have examples of problems and answers with illustrations. Careful review of the instructions and practice using example problems are the best ways to learn how to use the calculator.

Keep in mind that the calculator you select should be large enough so that you can use it. Many of the modern calculators have keys so small that it is almost impossible to hit just one key. You will be doing a considerable amount of work during this study effort—make it as easy on yourself as you can.

Another significant point to keep in mind when selecting a calculator is the importance of purchasing a unit that has the functions you need. Although a calculator with a lot of functions may look impressive, it can be complicated to use. Generally, the wastewater plant operator requires a calculator that can add, subtract, multiply, and divide. A calculator with parentheses functions is helpful, and, if you must calculate geometric means for fecal coliform reporting, logarithmic capability is helpful.

In many cases, calculators can be used to perform several mathematical functions in succession. Because various calculators are designed using different operating systems, you must review the instructions carefully to determine how to make the best use of the system.

Finally, it is important to keep a couple of basic rules in mind when performing calculations:

- Always write down the calculation you wish to perform.
- Remove any parentheses or brackets by performing the calculations inside first.

3.3 BASIC MATH

✓ *Note*: It is assumed that the reader has a fundamental knowledge of basic mathematical operations. Thus, the purpose of the following sections is to provide only a brief review of the mathematical concepts and applications frequently employed by wastewater treatment plant operations.

3.3.1 SEQUENCE OF OPERATIONS

✓ In a series of additions, the terms may be placed in any order and grouped in any way.

Examples:

$4 + 6 = 10$ and $6 + 4 = 10$

$(4 + 5) + (3 + 7) = 19$, $(3 + 5) + (4 + 7) = 19$, and $[7 + (5 + 4)] + 3 = 19$

✓ In a series of subtractions, changing the order or the grouping of the terms may change the result.

Examples:

$100 - 20 = 80$, but $20 - 100 = -80$

$(100 - 30) - 20 = 50$, but $100 - (30 - 20) = 90$

✓ When no grouping is given, the subtractions are performed in the order written—from left to right.

Examples:

$100 - 30 - 20 - 3 = 47$

or by steps, $100 - 30 = 70$, $70 - 20 = 50$, $50 - 3 = 47$

✓ In a series of multiplications, the factors may be placed in any order and in any grouping.

Examples:

$[(3 \times 3) \times 5] \times 6 = 270$ and $5 \times [3 \times (6 \times 3)] = 270$

✓ In a series of divisions, changing the order or the grouping may change the result.

Examples:

$100 \div 10 = 10$, but $10 \div 100 = 0.1$

$(100 \div 10) \div 2 = 5$, but $100 \div (10 \div 2) = 20$

✓ If no grouping is indicated, the divisions are performed in the order written—from left to right.

Examples:

$100 \div 5 \div 2$ is understood to mean $(100 \div 5) \div 2$

✓ In a series of mixed mathematical operations, the rule of thumb is, whenever no grouping is given, multiplications and divisions are to be performed in the order written, then additions and subtractions in the order written.

Let's look at a few example operations.

Example 3.1

$10 + 3 - 5 + 9 + 7 - 3 = 21$, by performing operations in the order in which they were given.

Example 3.2

$120 \div 3 \div 5 \times 2 \div 2 = 200$, by performing the operations from left to right—in the order they occur.

Example 3.3

$(12 \div 4) + (6 \times 2) - (6 \div 3) + (7 \times 3 \times 2) - 8 = ?$

$3 + 12 - 2 + 42 - 8 = 47$

First, perform the multiplications and divisions and then the additions and subtractions.

✓ It is important to group the indicated operations as indicated in example 3.4.

Example 3.4

In a series of different operations, parentheses ( ) and brackets [ ] can be used to group operations in the desired order. Thus, $120 \div 3 \times 5 \times 2 \div 2 = \{[(120 \div 3)5]2\} \div 2 = 200$.

3.3.1.1 Exercise 3.1

✓ *Note:* In problems 3.1 to 3.5, remember that the operations indicated by the *inmost groups* should be performed first.

3.1 $12 \div 3 + 8 \times 2 - 6 \div 2 + 7 = ?$

3.2 $14 + 16 - 3 + 10 - 4 - 6 = ?$

3.3 $16 \div 8 + 4 \times 2 \times 3 - 16 \times 2 \div 4 = ?$

3.4 $15 - 2 \times 3 - 15 \div 5 + 4 = ?$

3.5 $60 - 25 \div 5 + 15 - 100 \div 4 \times 2 = ?$

Note: Check Appendix A for answers.

3.3.2 CONVERSION OF FRACTIONS TO DECIMALS

A fraction is an incomplete division. With the availability of calculators, the easiest method for working with fractions is to convert the fraction into decimal form. To convert the fraction, divide the numerator (top number) by the denominator (bottom number).

$$\text{Fraction} = \frac{\text{Numerator}}{\text{Denominator}} \qquad (3.1)$$

Let's try a fraction to decimal conversion.

Example 3.5

What is the decimal equivalent of the fraction 5/8?

$$5/8 = 5 \div 8 = 0.625$$

Now let's convert the decimal to a common fraction.

Example 3.6

Convert 0.625 to a common fraction.

$$0.625 = \frac{625}{1000} = \frac{25}{40} = \frac{5}{8}$$

3.3.3 PERCENT

The words "per cent" mean "by the hundred." Percentage is often designated by the symbol %. Thus, 15% means 15 percent or 15/100 or 0.15. These equivalents may be written in the reverse order: 0.15 = 15/100 = 15%. In wastewater treatment, percent is frequently used to express plant performance and for control of sludge treatment processes.

✓ To determine percent, divide the quantity you wish to express as a percent by the total quantity then multiply by 100.

$$\text{Percent} = \frac{\text{Quantity} \times 100}{\text{Total}} \qquad (3.2)$$

Let's look at a couple of percent problems typically performed in wastewater treatment.

Example 3.7

The plant operator removes 6,500 gal of sludge from the settling tank. The sludge contains 325 gal of solids. What is the percent solids in the sludge?

$$\text{Percent} = \frac{325\,\text{gal}}{6{,}500\,\text{gal}} \times 100$$

$$= 5\%$$

Example 3.8

Convert 65% to decimal percent.

$$\text{Decimal percent} = \frac{\text{percent}}{100}$$

$$= \frac{65}{100}$$

$$= 0.65$$

Example 3.9

Sludge contains 5.8% solids. What is the concentration of solids in decimal percent?

$$\text{Decimal percent} = \frac{5.8\%}{100} = 0.058$$

✓ Unless otherwise noted, all calculations in the handbook using percent values require the percent be converted to a decimal before use.

✓ To determine what quantity a percent equals, first convert the percent to a decimal then multiply by the total quantity.

$$\text{Quantity} = \text{Total} \times \text{Decimal percent} \tag{3.3}$$

Example 3.10

Sludge drawn from the settling tank is 5% solids. If 2,800 gallons of sludge are withdrawn, how many gallons of solids are removed?

$$\text{Gallons} = \frac{5\%}{100} \times 2{,}800 \text{ gallons} = 140 \text{ gal}$$

3.3.4 ARITHMETIC AVERAGE (OR ARITHMETIC MEAN)

During the day-to-day operation of a treatment plant, much data are collected. These data, if properly evaluated, can provide useful information for trend analysis and can indicate how well the plant or unit process is operating. However, because there may be much variation in the data information, it is often difficult to determine trends in performance.

Arithmetic average refers to a statistical calculation used to describe a series of numbers, such as test results. By calculating an average, a group of data is represented by a single number. This number may be considered typical of the group. The arithmetic mean is the most commonly used measurement of average value.

✓ When evaluating information based on averages, remember that the "average" reflects the general nature of the group and does not necessarily reflect any one element of that group.

Arithmetic average is calculated by dividing the sum of all of the available data points (test results) by the number of test results.

$$\text{Average} = \frac{\text{Test 1} + \text{Test 2} + \text{Test 3} + \ldots + \text{Test } N}{\text{Number of Tests Performed } (N)} \tag{3.4}$$

Let's take a look at a couple of examples of how average is determined.

Example 3.11

Effluent BOD test results for the treatment plant during the month of August are shown below.

Test 1   22 mg/L
Test 2   33 mg/L
Test 3   21 mg/L
Test 4   13 mg/L

What is the average effluent BOD for the month of August?

$$\text{Average} = \frac{22\,mg/L + 33\,mg/L + 21\,mg/L + 13\,mg/L}{4} = 22.3\,mg/L$$

Example 3.12

For the primary influent flow, the following composite-sampled solids concentrations were recorded for the week:

Monday   310 mg/L SS
Tuesday   322 mg/L SS
Wednesday   305 mg/L SS
Thursday   326 mg/L SS
Friday   313 mg/L SS
Saturday   310 mg/L SS
Sunday   320 mg/L SS
Total   2,206 mg/L SS

$$\text{Average SS} = \frac{\text{Sum of All Measurements}}{\text{Number of Measurements Used}}$$

$$= \frac{2{,}206\,mg/L\,SS}{7}$$

$$= 315.1\,mg/L\,SS$$

3.3.5 POWERS OF TEN AND SCIENTIFIC NOTATION

✓ *Note:* In practice, the wastewater operator should realize that the accuracy of a final answer can never be better than the accuracy of the data used. Having stated the obvious, it should also be pointed out that correct and accurate data are worthless unless the operator is able to make correct computations.

Two common methods of expressing a number—powers of ten and scientific notation—will be discussed in this section.

3.3.5.1 Powers of Ten Notation

An expression such as 5^7 is a shorthand method of writing multiplication. For example, 5^7 can be written as

$$5 \times 5 \times 5 \times 5 \times 5 \times 5 \times 5$$

The expression 5^7 is referred to as 5 to the seventh power and is composed of an *exponent* and a *base number*.

✓ An exponent (or power of) indicates how many times a number is to be multiplied together. The base is the number being multiplied.

$$5^7 \leftarrow \text{(exponent)}$$
$$\text{(base)}$$

Let's look at base 5 again, using different exponents.

$$5^3$$

This is referred to as 5 to the third power, or 5 cubed. In expanded form,

$$5^3 = (5)(5)(5)$$

The expression 5 to the fourth power is written as

$$5^4$$

In expanded form, this notation means

$$5^4 = (5)(5)(5)(5)$$

These same considerations apply to letters (*a, b, x, y,* etc.) as well. For example:

$$z^2 = (z)(z) \quad \text{or} \quad z^4 = (z)(z)(z)(z)$$

✓ When a number or letter does not have an exponent, it is considered to have an exponent of one.

$$\text{Thus} \quad 5 = 5^1 \quad \text{or} \quad z = z^1$$

The following examples help to illustrate the concept of powers notation.

Example 3.13

How is the term 2^3 written in expanded form?
The power (exponent) of 3 means that the base number (2) is multiplied by itself three times:

$$2^3 = (2)(2)(2)$$

Example 3.14

How is the term in.^{2} written in expanded form? The power (exponent) of 2 means that the term (in.) is multiplied by itself two times:

Basic Math

$$in.^2 = (in.)(in.)$$

Example 3.15

How is the term 8^5 written in expanded form? The exponent of 5 indicates that 8 is multiplied by itself five times:

$$8^5 = (8)(8)(8)(8)(8)$$

Example 3.16

How is the term z^6 written in expanded form? The exponent of 6 indicates that z is multiplied by itself six times.

$$z^6 = (z)(z)(z)(z)(z)(z)$$

Example 3.17

How is the term $(3/8)^2$ written in expanded form?

✓ When parentheses are used, the exponent refers to the entire term within the parentheses. Thus, in this example, $(3/8)^2$ means

$$(3/8)^2 = (3/8)(3/8)$$

✓ When a negative exponent is used with a number or term, a number can be reexpressed using a positive exponent:

$$6^{-3} = 1/6^3$$

Another example is

$$11^{-5} = 1/11^5$$

Example 3.18

How is the term 8^{-3} written in expanded form?

$$8^{-3} = \frac{1}{8^3} = \frac{1}{(8)(8)(8)}$$

✓ *Note:* Any number or letter such as 3^0 or X^0 does not equal 3×1 or $X1$, but simply 1.

Example 3.19

When a term is given in expanded form, you can determine how it would be written in exponential form. For example,

$$(5)(5)(5) = 5^3$$

or

$$(\text{in.})(\text{in.}) = (\text{in.})^2$$

Example 3.20

Write the following term in exponential form:

$$\frac{(3)(3)}{(7)(7)(7)}$$

The exponent for the numerator (remember: numerator/denominator) of the fraction is 2 and the exponent for the denominator is 3. Therefore, the term would be written as

$$\frac{(3)^2}{(7)^3}$$

It should be pointed out that because the exponents are not the same, parentheses cannot be placed around the fraction and a single exponent cannot be used.

Example 3.21

✓ It is common to see powers used with a number or term used to denote area or volume units (in.^2, ft^2, in.^3, ft^3) and in scientific notation.

Write the following term in exponential form:

$$\frac{(\text{in.})(\text{in.})}{(\text{ft})(\text{ft})}$$

The exponent of both the numerator and denominator is 2:

$$\frac{(\text{in.})^2}{(\text{ft})^2}$$

The exponents are the same, which allows the use of parentheses to express the term as follows:

$$\left(\frac{\text{in.}}{\text{ft}}\right)^2$$

Example 3.22

✓ In moving a power from the numerator of a fraction to the denominator, or vice versa, the sign of the exponent is changed. For example

$$\frac{3^3 \times 4^{-2}}{8} = \frac{3^3}{8 \times 4^2}$$

3.3.5.2 Scientific Notation

✓ Scientific notation is a method by which any number can be expressed as a term multiplied by a power of 10. The term is always a term multiplied by a power of 10. The term is always greater than or equal to 2 but less than 10.

Examples of powers of 10 are

$$3.2 \times 10^1$$

$$1.8 \times 10^3$$

$$9.550 \times 10^4$$

$$5.31 \times 10^{-2}$$

✓ The numbers can be taken out of scientific notation by performing the indicated multiplication. For example,

$$3.2 \times 10^1 = (3.2)(10)$$
$$= 32$$
$$1.8 \times 10^3 = (1.8)(10)(10)(10)$$
$$= 1,800$$
$$9.550 \times 10^4 = (9.550)(10)(10)(10)(10)$$
$$= 95,500$$
$$5.31 \times 10^{-2} = (5.31) 1/10^2$$
$$= .0531$$

An easier way to take a number out of scientific notation is to move the decimal point the number of places indicated by the exponent.

RULE 1

Multiply by the power of 10 indicated. A positive exponent indicates a decimal move to the *right*, and a negative exponent indicates a decimal move to the *left*.

Example 3.23

Using the same examples above, the decimal point move rather than the multiplication method is performed as follows:

$$3.2 \times 10^1$$

The positive exponent of 1 indicates that the decimal point in 3.2 should be moved one place to the right:

$$3.2 = 32$$

The next example is

$$1.8 \times 10^3$$

The positive exponent of 3 indicates that the decimal point in 1.8 should be moved three places to the right:

$$1.800 = 1,800$$

The next example is

$$9.550 \times 10^4$$

The positive exponent of 4 indicates that the decimal point should be moved four places to the right:

$$9.5500 = 95,500$$

The final example is

$$5.31 \times 10^{-2}$$

The negative exponent of 2 indicates that the decimal point should be moved two places to the left:

$$05.31 = .0531$$

Example 3.24

Take the following number out of scientific notation:

$$3.516 \times 10^4$$

The positive exponent of 4 indicates that the decimal point should be moved 4 places to the right.

$$3.5160 = 35,160$$

Example 3.25

Take the following number out of scientific notation:

$$3.115 \times 10^{-4}$$

$$0003.115 = 0.0003115$$

✓ There are very few instances in which you will need to put a number or numbers into scientific notation, but you should know how to do it, if required. Thus, the method is discussed next. However, before demonstrating the process of putting a number into scientific notation, it is important to point out the procedure and the rule involved with the process.

Procedure:

When placing a number into scientific notation, place a decimal point after the first nonzero digit. (Remember that if no decimal point is shown in the number to be converted, it is assumed to be at the end of the number.) Count the number of places from the standard position to the original decimal point. This represents the exponent of the power of 10.

RULE 2

When a number is put into scientific notation, a decimal point move to the left indicates a positive exponent; a decimal point move to the right indicates a negative exponent.

Now let's try converting a few numbers into scientific notation.
First, convert 69 into scientific notation.

✓ *Note:* In order to obtain a number between 1 and 9, the decimal must be moved one place to the left. This move of one place gives the exponent, and the move to the left means that the exponent is positive:

$$69 = 6.9 \times 10^1$$

Let's try converting another number

$$1{,}500$$

✓ Remember, in order to obtain a number between 1 and 9, the decimal point must be moved three places to the left. The number of place moves (three) becomes the exponent of the power of 10, and the move to the left indicates a positive exponent:

$$1{,}500 = 1.5 \times 10^3$$

Let's try a decimal number

$$0.0661$$

$$0.0661 = 6.61 \times 10^{-2}$$

Example 3.26

Put the following number into scientific notation:

$$6{,}969{,}000$$

$$6{,}969{,}000 = 6.969 \times 10^6$$

Example 3.27

Convert the following decimal to scientific notation:

$$0.000696$$

$$0.000696 = 6.96 \times 10^{-4}$$

3.3.6 DIMENSIONAL ANALYSIS

Dimensional analysis is a valuable tool used to check if you have set up a problem correctly. In using dimensional analysis to check a math setup, you work with the dimensions (units of measure) only—not with numbers. In order to use the dimensional analysis method, you must know how to perform three basic operations:

✓ *Basic Operation:*

(1) To complete a division of units, always ensure that all units are written in the same format; it is best to express a horizontal fraction (such as gal/ft^3) as a vertical fraction.

$$\text{horizontal to vertical}$$

$$\text{gal/cu ft to } \frac{\text{gal}}{\text{cu ft}}$$

$$\text{psi to } \frac{\text{lb}}{\text{sq in.}}$$

Let's apply these procedures in the following examples shown below.

$$\text{ft}^3/\text{min becomes } \frac{\text{ft}^3}{\text{min}}$$

$$\text{s/min becomes } \frac{\text{s}}{\text{min}}$$

✓ *Basic Operation:*

(2) You must know how to divide by a fraction. Let's try dividing by a fraction. For example,

$$\frac{\frac{\text{lb}}{\text{d}}}{\frac{\text{min}}{\text{d}}} \text{ becomes } \frac{\text{lb}}{\text{d}} \times \frac{\text{d}}{\text{min}}$$

In the above problem, you may have noticed that the terms in the denominator were inverted before the fractions were multiplied. This is a standard rule that must be followed when dividing fractions.

Another example is

$$\frac{\text{mm}^2}{\frac{\text{mm}^2}{\text{m}^2}} \text{ becomes mm}^2 \times \frac{\text{m}^2}{\text{mm}^2}$$

✓ *Basic Operation:*

(3) You must know how to cancel or divide terms in the numerator and denominator of a fraction.

After fractions have been rewritten in the vertical form and division by the fraction has been reexpressed as multiplication as shown above, then the terms can be canceled (or divided) out.

✓ *Note:* For every term that is canceled in the numerator of a fraction, a similar term must be canceled in the denominator and vice versa, as shown below:

$$\frac{kg}{\cancel{d}} \times \frac{\cancel{d}}{min} = \frac{kg}{min}$$

$$\cancel{mm^2} \times \frac{m^2}{\cancel{mm^2}} = m^2$$

$$\frac{\cancel{gal}}{min} \times \frac{ft^3}{\cancel{gal}} = \frac{ft^3}{min}$$

Question: How are units that include exponents calculated?

When written with exponents, such as ft^3, a unit can be left as is or put in expanded form, (ft)(ft)(ft), depending on other units in the calculation. The point is that it is important to ensure that square and cubic terms are expressed uniformly, as sq ft, cu ft, or as ft^2, ft^3. For dimensional analysis, the latter system is preferred.

For example, let's say that you wish to convert 1,400 ft^3 volume to gallons, and you will use 7.48 gal/ft^3 in the conversion. The question becomes: Do you multiply or divide by 7.48?

In the above instance, it is possible to use dimensional analysis to answer this question; that is, are we to multiply or divide by 7.48?

In order to determine if the math setup is correct, only the dimensions are used.

First, try dividing the dimensions:

$$\frac{ft^3}{gal/ft^3} = \frac{ft^3}{\frac{gal}{ft^3}}$$

Then, the numerator and denominator are multiplied to get

$$= \frac{ft^6}{gal}$$

So, by dimensional analysis you determine that if you divide the two dimensions (ft^3 and gal/ft^3), the units of the answer are ft^6/gal, not gal. It is clear that division is not the right way to go in making this conversion.

What would have happened if you had multiplied the dimensions instead of dividing?

$$(ft^3)(gal/ft^3) = (ft^3)\left(\frac{gal}{ft^3}\right)$$

Then, multiply the numerator and denominator to obtain

$$= \frac{(ft^3)(gal)}{ft^3}$$

And cancel common terms to obtain

$$= \frac{(\cancel{ft^3})(gal)}{\cancel{ft^3}}$$

$$= gal$$

Obviously, by multiplying the two dimensions (ft^3 and gal/ft^3), the answer will be in gallons, which is what you want. Thus, because the math setup is correct, you would then multiply the numbers to obtain the number of gallons.

$$(1{,}400\ ft^3)(7.48\ gal/ft^3) = 10{,}472\ gal$$

Now let's try another problem with exponents. You wish to obtain an answer in square feet. If you are given the two terms—70 ft^3/s and 4.5 ft/s—is the following math setup correct?

$$(70\ ft^3/s)(4.5\ ft/s)$$

First, only the dimensions are used to determine if the math setup is correct. By multiplying the two dimensions, you get

$$(ft^3/s)(ft/s) = \left(\frac{ft^3}{s}\right)\left(\frac{ft}{s}\right)$$

Then multiply the terms in the numerators and denominators of the fraction:

$$= \frac{(ft^3)(ft)}{(s)(s)}$$

$$= \frac{ft^4}{s^2}$$

Obviously, the math setup is incorrect because the dimensions of the answer are not square feet. Therefore, if you multiply the numbers as shown above, the answer will be wrong.

Let's try division of the two dimensions instead.

$$ft^3/s = \frac{\dfrac{ft^3}{s}}{\dfrac{ft}{s}}$$

Invert the denominator and multiply to get

$$= \left(\frac{ft^3}{(s)}\right)\left(\frac{s}{(ft)}\right)$$

$$= \frac{(ft)(ft)(ft)(s)}{(s)(ft)}$$

$$= \frac{(ft)(ft)(ft)(\cancel{s})}{(\cancel{s})(\cancel{ft})}$$

$$= ft^2$$

Because the dimensions of the answer are square feet, this math setup is correct. Therefore, by dividing the numbers as was done with units, the answer will also be correct.

$$\frac{70 \text{ ft}^3/s}{4.5 \text{ ft}/s} = 15.56 \text{ ft}^2$$

Example 3.28

You are given two terms 5 m/s and 7 m²—and the answer to be obtained is in cubic meters per second (m³/s). Is multiplying the two terms the correct math setup?

$$(m/s)(m^2) = \frac{m}{s} \times m^2$$

Multiply the numerators and denominator of the fraction:

$$= \frac{(m)(m^2)}{s}$$

$$= \frac{m^3}{s}$$

Because the dimensions of the answer are cubic meters per second (m³/s), the math setup is correct. Therefore, multiply the numbers to get the correct answer.

$$5(m/s)(7 \text{ m}^2) = 35 \text{ m}^3/s$$

Example 3.29

Solve the following problem:
Given: The flow rate in a water line is 2.3 ft³/s. What is the flow rate expressed as gallons per minute?
Set up the math problem and then use dimensional analysis to check the math setup:

$$(2.3 \text{ ft}^3/s)(7.48 \text{ gal}/\text{ft}^3)(60 \text{ s}/\text{min})$$

Dimensional analysis is used to check the math setup:

$$(\text{ft}^3/s)(\text{gal}/\text{ft}^3)(s/\text{min}) = \left(\frac{\text{ft}^3}{s}\right)\left(\frac{\text{gal}}{\text{ft}^3}\right)\left(\frac{s}{\text{min}}\right)$$

$$= \frac{\cancel{\text{ft}^3}}{\cancel{s}} \frac{\text{gal}}{\cancel{\text{ft}^3}} \frac{\cancel{s}}{\text{min}}$$

$$= \frac{\text{gal}}{\text{min}}$$

The math setup is correct as shown above. Therefore, this problem can be multiplied out to get the answer in correct units

$$(2.3 \text{ ft}^3/\text{s})(7.48 \text{ gal}/\text{ft}^3)(60 \text{ s}/\text{min}) = 1032.24 \text{ gal}/\text{min}$$

3.3.7 ROUNDING OFF A NUMBER

It is sometimes necessary to round off measurements to a certain number of significant figures. The number of significant figures in a measurement is the number of digits that are known for sure, plus one more digit that is an estimate. The number 6.93436 cm, for example, has six significant figures. For numbers less than one, such as 0.00696, the zeros to the right of the decimal point are not considered significant figures; thus, 0.00696 has three significant figures. For numbers greater than 10, such as 696,000, the zeros should not be considered significant; thus, the number 696,000 has three significant figures.

Rounding off a number means replacing the final digit of a number with zeros, thus expressing the number as tens, hundreds, thousands, ten thousands, or tenths, hundredths, thousandths, ten thousandths, etc. (e.g., 498 as 500; 0.49 as 0.5; or 6,696,696 as 7,000,000).

✓ There is a basic rule to be followed when rounding off numbers.

RULE 3

A number is rounded off by dropping one or more numbers from the right and adding zeros if necessary to place the decimal point. If the last figure dropped is 5 or more, increase the last retained figure by 1. If the last figure dropped is less than 5, do not increase the last retained figure.

Example 3.30

Rounding to significant figures: Round off 11,547 to four, three, two, and one significant figure.

Solution:

$$11,547 = 11,550 \text{ to four significant figures}$$

$$11,547 = 11,500 \text{ to three significant figures}$$

$$11,547 = 12,000 \text{ to two significant figures}$$

$$11,547 = 10,000 \text{ to one significant figure}$$

Example 3.31

Rounding to a particular place value in the decimal system: Round 47,937 to the nearest hundred. The procedure used in this rounding depends on the digit just to the right of the hundreds place:

$$47,937$$
(hundreds place)

Because the digit to the right of the hundreds place is less than 5, 9 is not changed, and all the digits to the right of 9 are replaced with zeros: 47,937 becomes 47,900 (rounded to the nearest hundred).

Let's look at another example where a decimal number is to be rounded.
Round 6.653 to the nearest tenth.

$$6.6\underset{\uparrow}{5}3$$
(tenths place)

The digit to the right of the tenths place is 5. Therefore, the 6 is increased by 1, and all digits to the right are dropped: 6.653 becomes 6.7 (rounded to the nearest tenth).

3.3.8 EQUATIONS: SOLVING FOR THE UNKNOWN

In wastewater, operations related to calculations used in process operations may require you to use equations to solve for the unknown quantity. To make these calculations, you must first know the values for all but one of the terms of the equation to be used.

✓ An *equation* is a statement that two expressions or quantities are equal in value. The statement of equality $6x + 4 = 19$ is an equation; that is, it is algebraic shorthand for "The sum of 6 times a number plus 4 is equal to 19." It can be seen that the equation $6x + 4 = 19$ is much easier to work with than the equivalent sentence (for some folks).

When thinking about equations, it is helpful to consider an equation as being similar to a balance. The equal sign tells you that two quantities are "in balance" (i.e., they are equal).

Let's get back to the equation $6x + 4 = 19$. The solution to this problem may be summarized in three steps.

Step (1) $6x + 4 = 19$
Step (2) $6x = 15$
Step (3) $x = 2.5$

✓ Step 1 expresses the whole equation. In Step 2, 4 has been subtracted from both members of the equation. In Step 3, both members have been divided by 6.

✓ An equation is, therefore, kept in balance (both sides of the equal sign are kept equal) by subtracting the same number from both members (sides), adding the same number to both, or dividing or multiplying by the same number.

The expression $6x + 4 = 19$ is called a *conditional equation,* because it is true only when x has a certain value. The number to be found in a conditional equation is called the *unknown number* the *unknown quantity* or, more briefly, the *unknown.*

✓ Solving an equation is finding the value or values of the unknown that make the equation true.

Let's take a look at another equation:

$$W = F \times D \tag{3.5}$$

where

W = work

F = force

D = distance

Thus,

$$\text{Work} = \text{Force (lb)} \times \text{Distance (ft or in.)}$$

$$= \text{ft-lb or in-lb}$$

Suppose you have this equation:

$$60 = (x)(2)$$

How can you determine the value of *x*? By following the axioms presented in Section 3.3.8.1, the solution to the unknown is quite simple.

✓ It is important to point out that the following discussion includes only what the axioms are and how they work. (If you are interested in why these axioms work and how they came about, you should consult an algebra text.)

3.3.8.1 Axioms

(1) If equal numbers are added to equal numbers, the sums are equal.
(2) If equal numbers are subtracted from equal numbers, the remainders are equal.
(3) If equal numbers are multiplied by equal numbers, the products are equal.
(4) If equal numbers are divided by equal numbers (except zero), the quotients are equal.
(5) Numbers that are equal to the same number or to equal numbers are equal to each other.
(6) Like powers of equal numbers are equal.
(7) Like roots of equal numbers are equal.
(8) The whole of anything equals the sum of all its parts.

Note: Axioms 2 and 4 were used to solve the equation $6x + 4 = 19$.

3.3.8.2 Solving Simple Equations

✓ As stated earlier, solving an equation is determining the value or values of the unknown number or numbers in the equation.

Example 3.32

Find the value of *x* if $x - 8 = 2$.

Here you can see by inspection that $x = 10$, but inspection does not help in solving more complicated equations. However, if you notice that to determine that $x = 10$, 8 is added to each member of the given equation, you have acquired a method or procedure that can be applied to similar but more complex problems.

Solution: Given equation:

$$x - 8 = 2$$

Add 8 to each member (axiom 1),

$$x = 2 + 8$$

Collecting the terms (that is, adding 2 and 8),

$$x = 10$$

Example 3.33

Solve for *x*, if $4x - 4 = 8$ (each side is in simplest terms)

Solution:

$4x = 8 + 4$ [the term (-4) is moved to the right of the equal sign as $(+4)$]

$4x = 12$

$$\frac{4x}{4} = \frac{12}{4} \text{ (divide both sides)}$$

$x = 3$ (we're done; *x* is alone on the left and is equal to the value on the right)

Example 3.34

Solve for *x*, if $x + 10 = 15$

Solution:

Subtract 10 from each member (axiom 2),

$$x = 15 - 10$$

Collect the terms,

$$x = 5$$

Example 3.35

Solve for *x*, if $5x + 5 - 7 = 3x + 6$

Solution:

Collect the terms $(+5)$ and (-7):

$$5x - 2 = 3x + 6$$

Add 2 to both members (axiom 1):

$$5x = 3x + 8$$

Subtract $3x$ from both members (axiom 2)

$$2x = 8$$

Divide both members by 2 (axiom 4):

$$x = 4$$

3.3.8.3 Checking the Answer

After you have obtained a solution to an equation, you should always check it. This is an easy process. All you need to do is substitute the solution for the unknown quantity in the given equation. If the two members of the equation are then identical, the number substituted is the correct answer.

Example 3.36

Solve and check $4x + 5 - 7 = 2x + 6$

Solution:

$$4x + 5 - 7 = 2x + 6$$
$$4x - 2 = 2x + 6$$
$$4x = 2x + 8$$
$$2x = 8$$
$$x = 4$$

Substituting the answer $x = 4$ in the original equation,

$$4x + 5 - 7 = 2x + 6$$
$$4(4) + 5 - 7 = 2(4) + 6$$
$$16 + 5 - 7 = 8 + 6$$
$$14 = 14$$

Because the statement $14 = 14$ is true, the answer $x = 4$ must be correct.

3.3.8.4 Setting up Equations

The equations discussed in the preceding paragraphs were expressed in *algebraic* language. It is important to learn how to set up an equation by translating a sentence into an equation (into algebraic language) and then solving this equation.

✓ The following suggestions and examples should help you:

(1) Always read the statement of the problem carefully.
(2) Select the unknown number and represent it by some letter. If more than one unknown quantity exists in the problem, try to represent those numbers in terms of the same letter—that is, in terms of one quantity.
(3) Develop the equation, using the letter or letters selected, and then solve.

Example 3.37

Given: One number is eight more than another. The larger number is two less than three times the smaller. What are the two numbers?

Solution: Let *n* represent the small number.

Then *n* + 8 must represent the larger number.
$$n + 8 = 3n - 2$$
$$n = 5 \text{ (small number)}$$
$$n + 8 = 13 \text{ (large number)}$$

Example 3.38

Given: If five times the sum of a number and 6 is increased by 3, the result is two less than 10 times the number. Find the number.

Solution: Let *n* represent the number.
$$5(n + 6) + 3 = 10n - 2$$
$$n = 5$$

3.3.9 RATIO AND PROPORTION

3.3.9.1 Ratio

✓ Ratio is the comparison of two numbers by division or an indicated division. The ratio of one number to another is determined when the one number is divided by the other.

A ratio always includes two numbers. For example: If a box has a length and width of 8 in. and 4 in., the ratio of the length to the width is expressed as 8/4 or 8:4. Both expressions have the same meaning.

All ratios are reduced to the lowest possible terms. This is similar to reducing a fraction to the lowest possible terms. The ratio of 8/4 or 8:4 should be reduced to its lowest possible terms by dividing the 8 and the 4 by 4. The resulting ratio is 2:1 or 2/1.

The ratio of the length of the box to its width is 2:1, because the box is two times as long as it is wide.

This ratio can also be stated as the relationship of the width to the length. The box is 4 in. wide and 8 in. long. The ratio of the width to the length is 4:8 or 4/8. This ratio when reduced becomes 1:2 or 1/2. The width of the box is 1/2 its length.

3.3.9.2 Proportions

✓ Simply put, a proportion is a statement of equality between two ratios. Thus, 2:4 = 4:8 is a proportion. It is evident that the two ratios are equal when the proportion is written in fractional form: 2/4 = 4/8. Either form may be read as follows: "Two is to four as four is to eight."

A general statement of the preceding proportion would be

$$\frac{a}{b} = \frac{c}{d} \text{ (fractional form)}$$

$$a{:}b = c{:}d \text{ (proportional form)}$$

where *a*, *b*, *c*, and *d* represent numbers.

A proportion may be written with a double colon (::) in place of the equals sign:

$$a:b::c:d$$

The first and last terms are called the extremes; the second and third terms are called the means. Thus, in the preceding examples, the a and d, are the extremes; the b and c, are the means.

By consideration of proportions as fractions, it becomes evident that if any three or four members are known, then the fourth can be determined. When the proportion given earlier, $a:b = c:d$, is written as a fraction, $a/b = c/d$, then $a \times d = c \times b$. If you substitute numbers for letters and for the proportion, then $1/2 = 2/4$, $2 \times 2 = 4$, and $1 \times 4 = 4$. It is, therefore, obvious that the product of the means (2×2) equals the product of the extremes (1×4).

$$\overset{\text{extremes}}{a{:}b = c{:}d}\underset{\text{means}}{} \qquad \overset{\text{extremes}}{1{:}2 = 2{:}4}\underset{\text{means}}{}$$

Example 3.39

What is x in the proportion $2:3 = x:12$?

Solution: Rewriting in fractional form

$$\frac{2}{3} = \frac{x}{12}$$

Because $12 = 4 \times 3$, x must be 2×4, or 8, because this gives equal fractions. The proportion when completed is

$$\frac{2}{3} = \frac{8}{12}$$

Another method of making use of the principle that the product of the means equals the product of the extremes is shown below:

$$2:3 = x:12$$

The product of the means, $3x$, equals the product of the extremes, 2×12 or 24. Thus, $3x = 24$, and $1x$ must equal 8, the same answer obtained earlier.

Example 3.40

If a pump will fill a tank in 20 hours at 4 gpm (gallons per minute), how long will it take a 10-gpm pump to fill the same tank?

First, analyze the problem. Here, the unknown is some number of hours. But should the answer be larger or smaller than 20 hours? If a 4-gpm pump can fill the tank in 20 hours, a larger pump (10-gpm) should be able to complete the filling in less than 20 hours. Therefore, the answer should be less than 20 hours.

Now set up the proportion:

$$\frac{x \text{ h}}{20 \text{ h}} = \frac{4 \text{ gpm}}{10 \text{ gpm}}$$

$$x = \frac{(4)(20)}{10}$$

$$x = 8 \text{ h}$$

It doesn't take long before you will gain an understanding of proportion problems that will allow you to skip some of the various steps to solving these problems (practice makes perfect and repetition aids easy recognition). In the following examples, a short-cut method is shown that will allow an experienced operator to solve problems quite easily.

Example 3.41

To make a certain chemical solution, 66.3 mg of chemical must be added to 150 L of water. How much of the chemical should be added to 25 L to make up the same strength solution?

To solve this problem, you must first decide what is unknown and whether you expect the unknown value to be larger or smaller than the known value of the same unit. The amount of chemical to be added to 25 L is the unknown, and you would expect this to be smaller than the 66.3 mg needed for 150 L.

First, take the two known quantities of the same unit (25 L and 150 L), and make a fraction to multiply the third known quantity (66.3 mg) by. Notice that there are two possible fractions you can make with 25 and 150:

$$\frac{25}{150} \text{ or } \frac{150}{25}$$

Second, you want to choose the fraction that will make the unknown number of milligrams less than the known (66.3 mg). Multiplying the 66.3 by the fraction 25/150 would result in a number smaller than 66.3. Multiplying 66.3 by 150/25, however, would result in a number larger than 66.3.

You wish to obtain a number smaller than 66.3, so multiply by 25/150; then complete the calculation to solve the problem:

$$\frac{25}{150}(66.3) = z$$

$$\frac{(25)(66.3)}{150} = z$$

$$11.05 \text{ mg} = z$$

From the above operation, it should be obvious that the key to this method is arranging the two known values of like units into a fraction that, when multiplied by the third known value, will render a result that is smaller or larger as required.

3.4 UNITS OF MEASUREMENT

A basic knowledge of units of measurement and how to use them is essential. Wastewater practitioners should be familiar both with the U.S. Customary System (USCS) or English System and the International System of Units (SI). Some of the important units are summarized in Table 3.1. Table 3.1 gives conversion factors between SI and USCS systems for some of the basic units that will be encountered.

In the study of wastewater treatment plant operations (and in actual practice), it is quite common

TABLE 3.1. Commonly Used Units and Conversion Factors.

Quantity	SI Units	SI Symbol	× Conversion Factor =	USCS Units
Length	meter	m	3.2808	ft
Mass	kilogram	kg	2.2046	lb
Temperature	Celsius	°C	1.8 (°C) + 32	°F
Area	square meter	m^2	10.7639	ft^2
Volume	cubic meter	m^3	35.3147	ft^3
Energy	kilojoule	Kj	0.9478	Btu
Power	watt	W	3.4121	Btu/hr
Velocity	meter/second	m/s	2.2369	mi/hr

to encounter both extremely large quantities and extremely small ones. The concentration of some toxic substance may be measured in parts per million or billion (ppm or ppb), for example.

✓ Remember: ppm may be roughly described as an amount contained in a shot glass in the bottom of a swimming pool.

To describe quantities that may take on such large or small values, it is useful to have a system of prefixes that accompanies the units. Some of the more important prefixes are presented in Table 3.2.

3.5 CIRCUMFERENCE, AREAS, AND VOLUMES

Wastewater treatment plants consist of a series of tanks and channels. Proper operational control requires that you be able to perform several process control calculations. Many of these calculations include parameters such as the circumference or perimeter, area, or the volume of the tank or channel as part of the information necessary to determine the result. To aid in performing these calculations, the following definitions are provided.

- *Area*   the area of an object, measured in square units.
- *Base*   the term used to identify the bottom leg of a triangle, measured in linear units.
- *Circumference*   the distance around an object, measured in linear units. When determined for other than circles, it may be called the perimeter of the figure, object, or landscape.
- *Cubic units*   measurements used to express volume, cubic feet, cubic meters, etc.
- *Depth*   the vertical distance from the bottom of the tank to the top. Normally measured in terms of liquid depth and given in terms of side wall depth (SWD), measured in linear units.

TABLE 3.2. Common Prefixes.

Quantity	Prefix	Symbol
10^{-12}	pico	p
10^{-9}	nano	n
10^{-6}	micro	µ
10^{-3}	milli	m
10^{-2}	centi	c
10^{-1}	deci	d
10	deca	da
10^2	hecto	h
10^3	kilo	k
10^6	mega	M

- *Diameter* the distance from one edge of a circle to the opposite edge passing through the center, measured in linear units.
- *Height* the vertical distance from the base or bottom of a unit to the top or surface.
- *Length* the distance from one end of an object to the other, measured in linear units.
- *Linear units* measurements used to express distances: feet, inches, meters, yards, etc.
- *Pi, π* a number in the calculations involving circles, spheres, or cones (π = 3.14).
- *Radius* the distance from the center of a circle to the edge, measured in linear units.
- *Sphere* a container shaped like a ball.
- *Square units* measurements used to express area, square feet, square meters, acres, etc.
- *Volume* the capacity of the unit, how much it will hold, measured in cubic units (cubic feet, cubic meters) or in liquid volume units (gallons, liters, million gallons).
- *Width* the distance from one side of the tank to the other, measured in linear units.

3.5.1 PERIMETER AND CIRCUMFERENCE

As stated previously, wastewater plant operators may be called upon to make certain measurements in the field related to process control calculations. Moreover, on occasion, it may be necessary to determine the distance around grounds or landscapes. In order to measure the distance around property, buildings, and basin-like structures, it is necessary to determine either perimeter or circumference. The *perimeter* is how far it is around an object, like a piece of ground. *Circumference* is the distance around a circle or circular object, such as a clarifier. Distance is linear measurement, which defines the distance (or length) along a line. Standard units of measurement like inches, feet, yards, and miles and metric units like centimeters, meters, and kilometers are used.

✓ The perimeter of a rectangle (a 4-sided figure with 4 right angles) is obtained by adding the lengths of the four sides.

$$\text{Perimeter} = L_1 + L_2 + L_3 + L_4 \qquad (3.6)$$

Find the perimeter of the following rectangle:

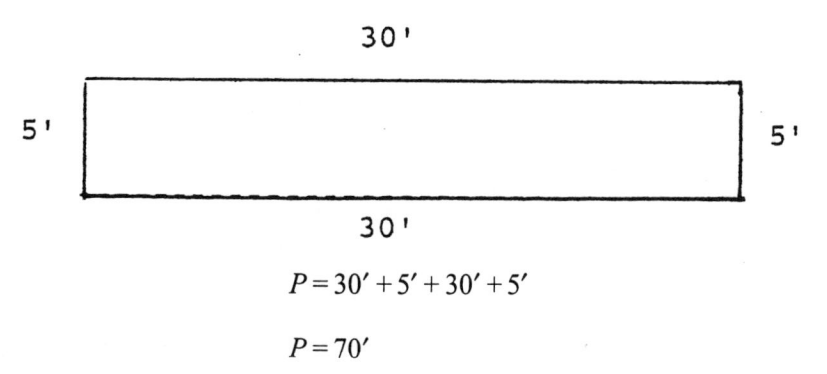

$$P = 30' + 5' + 30' + 5'$$

$$P = 70'$$

Example 3.42

You wish to install a fence around your plant site. It has the dimensions shown below. Determine the perimeter of the area so you can order the correct length of chain link fencing material.

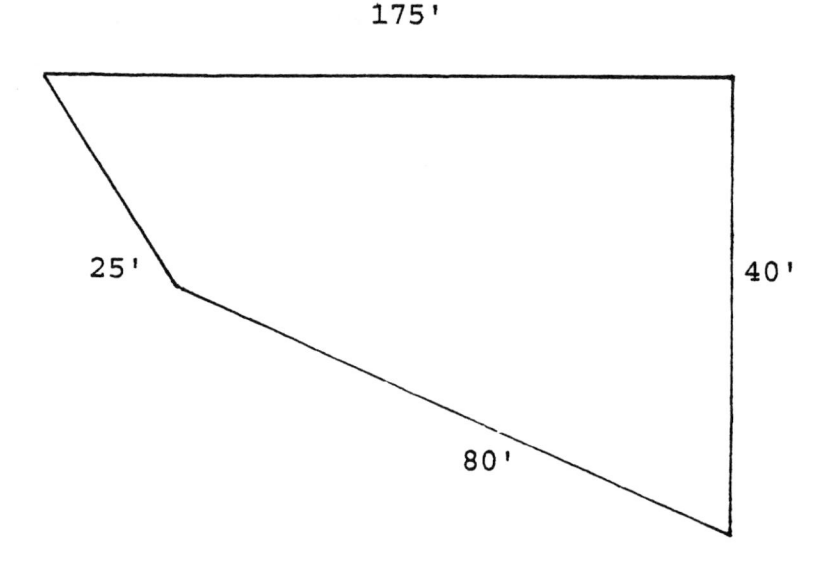

To calculate the perimeter, or distance around the area, add the lengths of all four sides:

$$P = 25' + 175' + 40' + 180'$$

$$P = 420'$$

✓ The circumference (*C*) of a circle is found by multiplying pi (π) times the diameter (*D*) (diameter is a straight line passing through the center of a circle—the distance across the circle).

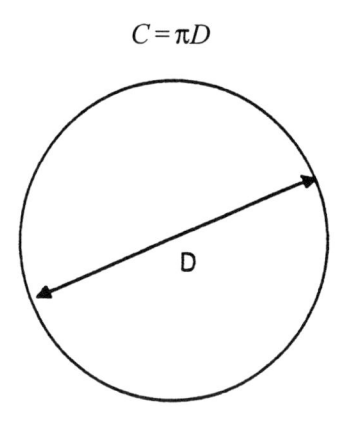

$$C = \pi D$$

where

C = circumference

π = Greek letter pi

π = 3.1416

D = diameter

Use this calculation if, for example, you must determine the circumference of a circular tank.

Example 3.43

Find the circumference of a circle that has a diameter of 25′ ($\pi = 3.14$).

$$C = \pi \times 25'$$

$$C = 3.14 \times 25'$$

$$C = 78.5 \text{ ft}$$

Example 3.44

A circular chemical holding tank has a diameter of 18 m. What is the circumference of this tank?

$$C = \pi(D)$$

$$C = (3.14)(\text{diameter})$$

$$C = (3.14)(18 \text{ m})$$

$$C = 56.52 \text{ m}$$

Example 3.45

An influent pipe inlet opening has a diameter of 6′. What is the circumference of the inlet opening in inches?

$$C = \pi(D)$$

$$C = 3.14 \times 6 \text{ ft}$$

$$C = 18.84 \text{ ft}$$

3.5.2 AREA

For area measurements in wastewater work, three basic shapes are particularly important, namely circles, rectangles, and triangles.

✓ *Area* is the amount of surface an object contains or the amount of material it takes to cover the surface. The area on top of a chemical tank is called the *surface area*. The area of the end of a ventilation duct is called the *cross-sectional area* (the area at right angles to the length of ducting). Area is usually expressed in square units, such as square inches (in.^{2}) or square feet (ft^2). Land may also be expressed in terms of square miles (sections) or acres (43,560 ft^2) or in the metric system as *hectares*.

The area of a rectangle is found by multiplying the length (L) times width (W).

$$\text{Area} = L \times W \qquad (3.7)$$

Example 3.46

Find the area of the following rectangle:

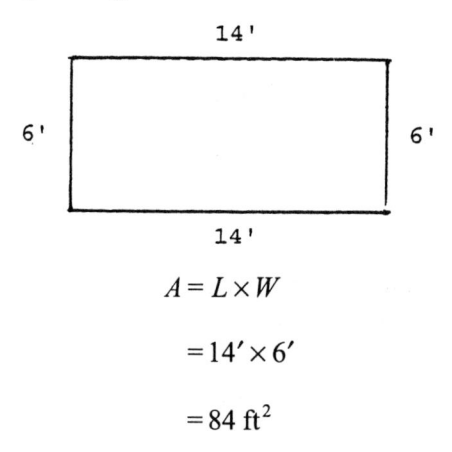

$$A = L \times W$$
$$= 14' \times 6'$$
$$= 84 \text{ ft}^2$$

✓ The surface area of a circle is determined by multiplying π times the radius squared.

✓ *Note:* Radius, designated *r*, is defined as a line from the center of a circle or sphere to the circumference of the circle or surface of the sphere.

$$\text{Area of circle} = \pi r^2$$

where

A = area

π = Greek letter pi (π = 3.14)

r = radius of a circle—radius is one-half the diameter

Example 3.47

What is the area of the circle shown below?

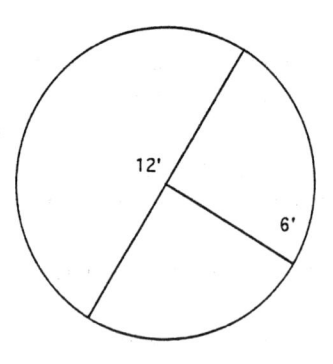

$$\text{Area of circle} = \pi r^2$$
$$= \pi 6^2$$
$$= 3.14 \times 36$$
$$= 113 \text{ ft}^2$$

Example 3.48

What is the area of the rectangle shown below?

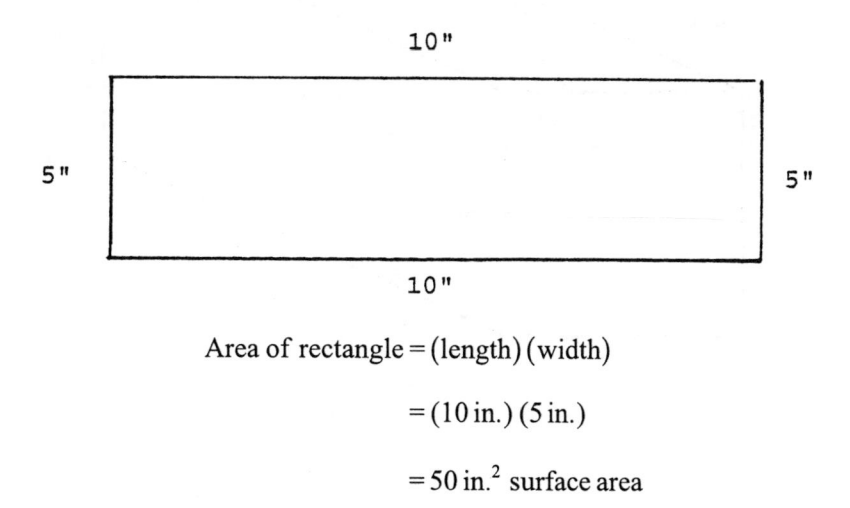

```
         10"
    ┌─────────────┐
 5" │             │ 5"
    └─────────────┘
         10"
```

Area of rectangle = (length)(width)

= (10 in.)(5 in.)

= 50 in.^{2} surface area

✓ *Note:* It should be pointed out that even though area measurements are expressed in square units, this does *not* mean that the surface must be square in order to measure it. The point is that the surface of virtually any shape can be measured.

3.5.3 VOLUME

✓ The amount of space occupied by or contained in an object, *volume,* is expressed in cubic units, such as cubic inches (in.^{3}), cubic feet (ft^3), acre feet (1 acre foot = 43,560 ft^3), etc.

The volume of a rectangular object is obtained by multiplying the length times the width times the depth or height.

$$V = L \times W \times H \qquad (3.8)$$

where

L = length

W = width

D or H = depth or height

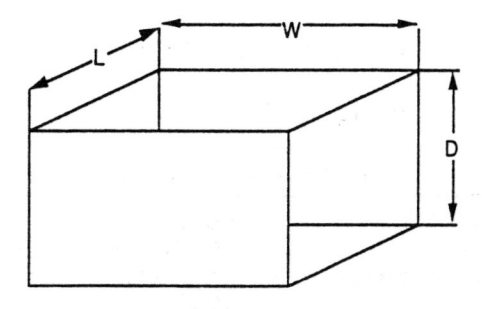

Example 3.49

Find the volume in cubic feet of a holding pond with the following dimensions:

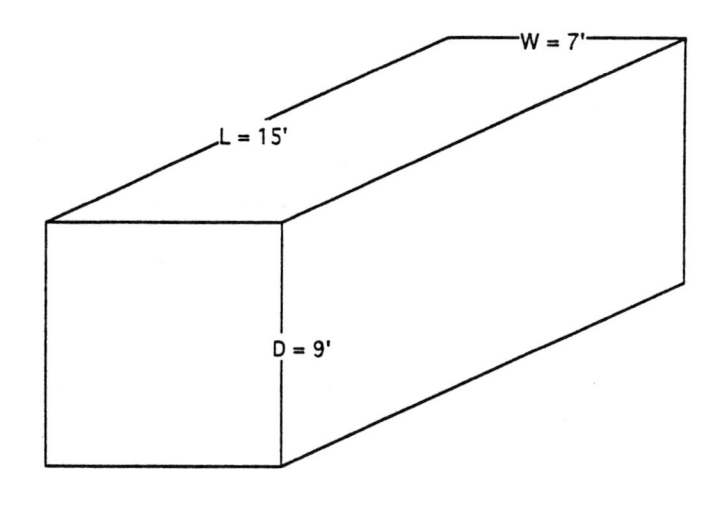

$$V = L \times W \times D$$
$$= 15' \times 7' \times 9'$$
$$= 945 \text{ ft}^3$$

For wastewater operators, representative surface areas are most often rectangles, triangles, circles, or a combination of these. Practical volume formulas used in wastewater calculations are given in Table 3.3.

Example 3.50

Find the volume of a 3-in. round pipe that is 300-ft long.

(1) Step 1: Change the diameter of the duct from inches to feet by dividing by 12.
$$D = 3 \div 12 = 0.25 \text{ ft}$$

(2) Step 2: Find the radius by dividing the diameter by 2.
$$r = 0.25 \text{ ft} \div 2 = 0.125$$

(3) Step 3: Find the volume.
$$V = L \times \pi r^2$$
$$V = 300 \text{ ft} \times \pi \times (0.0156) \text{ ft}^2$$
$$V = 14.70 \text{ ft}^2$$

TABLE 3.3. Volume Formulas.

Sphere volume	=	$(\pi/6)(\text{diameter})^3$
Cone volume	=	1/3 (volume of a cylinder)
Rectangular tank volume	=	(area of rectangle) (*D* or *H*)
	=	(*LW*) (*D* or *H*)
Cylinder volume	=	(area of cylinder) (*D* or *H*)
	=	πr^2 (*D* or *H*)
Triangle volume	=	(area of triangle) (*D* or *H*)
	=	(*bh*/2) (*D* or *H*)

Example 3.51

Find the volume of a smoke stack that is 24 in. in diameter (entire length) and 96 in. tall. Find the radius of the stack. The radius is one-half the diameter.

$$24 \text{ in.} \div 2 = 12 \text{ in.}$$

Find the volume.

$$V = H \times \pi r^2$$

$$V = 96 \text{ in.} \times \pi (12 \text{ in.})^2$$

$$V = 96 \text{ in.} \times \pi (144 \text{ in.}^2)$$

$$V = 43{,}407 \text{ ft}^3$$

3.5.3.1 Volume of a Cone and Sphere

3.5.3.1.1 Volume of a Cone

$$\text{Volume of cone} = \frac{\pi}{12} \times \text{Diameter} \times \text{Diameter} \times \text{Height} \tag{3.9}$$

✓
$$\frac{\pi}{12} = \frac{3.14}{12} = 0.262$$

✓ The diameter used in the formula is the diameter of the base of the cone.

Example 3.52

The bottom section of a circular settling tank has the shape of a cone. How many cubic feet of water are contained in this section of the tank if the tank has a diameter of 120 ft and the cone portion of the unit has a depth of 6 ft?

$$\text{Volume, ft}^3 = 0.262 \times 120 \text{ ft} \times 120 \text{ ft} \times 6 \text{ ft} = 22{,}637 \text{ ft}^3$$

3.5.3.1.2 Volume of a Sphere

$$\text{Volume} = \frac{3.14}{6} \times \text{Diameter} \times \text{Diameter} \times \text{Diameter} \tag{3.10}$$

✓
$$\frac{\pi}{6} = \frac{3.14}{6} = 0.524$$

Example 3.53

What is the volume of cubic feet of a gas storage container that is spherical and has a diameter of 60 ft?

$$\text{Volume, ft}^3 = 0.524 \times 60 \text{ ft} \times 60 \text{ ft} \times 60 \text{ ft} = 113{,}184 \text{ ft}^3 \tag{3.11}$$

3.6 GEOMETRIC MEAN

Geometric mean (a.k.a. geometric average) is a statistical calculation used for reporting bacteriological test results. It is not affected by wide shifts in test results to the same extent the arithmetic mean is affected. It can be computed using logarithms or by determining the Nth root of the product of the individual test results.

Although it is possible to perform each of the calculations without one, it is best to use a calculator that can perform logarithm (log) functions and/or exponential (Y^x) functions.

3.6.1 LOGARITHM (LOG) METHOD

To perform the calculations required to obtain a geometric mean using the log method, you must have a calculator capable of converting test results into their equivalent logarithms and converting the logarithm of the geometric mean back into its equivalent number *(antilog)*.

Procedure:

- Enter each test result into the calculator, and obtain its equivalent log value. Replace any zero (0) test result with a one (1) and determine the log of 1.
- Add all log values.
- Divide by the number of tests performed.
- Determine the antilog of this answer (the numerical equivalent of the log). The antilog is the geometric mean.

3.6.2 *N*th ROOT METHOD

The Nth root method for determination of the geometric mean requires a calculator that can multiply all of the test results together and then determine the Nth root of this number.

Procedure:

- Replace any zero (0) test result with a one (1).
- Multiply all of the reported test values (test 1 × test 2 × test 3 × ... × test N).
- Using the Nth root function (Y^x) of the calculator, determine the Nth root of the product obtained in the previous step.

Let's take a look at an example.

Example 3.54

Problem:

The results of the fecal coliform testing performed during the month of June are shown below. What is the geometric mean of the test results computed by the log method and the Nth root method?

Test 1	20
Test 2	0
Test 3	180
Test 4	2,133
Test 5	69
Test 6	96
Test 7	19
Test 8	44

Solution:

Step 1: Geometric Mean by the Log Method

		Log
Test 1	20	1.30103
Test 2	1	0.00000
Test 3	180	2.25527
Test 4	2,133	3.32899
Test 5	69	1.83884
Test 6	96	1.98227
Test 7	19	1.27875
Test 8	44	1.64345
Sum		13.62860

$$\text{Log of Geometric Means} = \frac{13.62860}{8} = 1.703575$$

Antilog of 1.703575 = 50.5 or 51

Step 2: Geometric Mean by the *N*th Root Method
Calculate mean by the *N*th root method. Calculate the product of all the test results during the period.

$$20 \times 1 \times 180 \times 2{,}133 \times 69 \times 96 \times 19 \times 44$$

Using the calculator, determine the 8th root of this number (8th because there are eight test results).

$$\text{8th Root} = 51$$

3.7 MOVING AVERAGE

Conducting and establishing trend analysis for use in process control and performance evaluation is important in wastewater treatment operations. To aid in this effort, the *moving average* computation is commonly used (it provides a method to develop trends for use in both process control and performance evaluation).

The moving average takes all available data into account, provides a leveling of erratic data points, and limits the length of time an individual data point will impact the computation. The moving average can be determined as an arithmetic or geometric mean and for varying periods (5, 7, or 28 days). The most common moving average is the seven-day arithmetic moving average. Because it is the period most commonly used in wastewater treatment, we describe the procedure for calculation of the seven-day arithmetic moving average in this section.

✓ It is important to note that a moving average can be calculated each day following completion of the initial data collection period (5, 7, or 28 days). Each day's moving average is calculated in the same way using the most recent data period.

Procedure:

- Add all the results of tests performed during the period Day 1 to Day 7.
- Divide the number of tests performed during this period.
- This is the seven-day moving average for Day 7.

- Repeat the procedure on Day 8 using the test results collected during the period Day 2 to Day 8. The result of this calculation is the moving average for Day 8.
- The same technique applies to all moving averages, only the days included in the calculation change.

$$\text{Moving Average} = \frac{\text{Test 1} + \text{Test 2} + \text{Test 3} + \text{Test 4} + \ldots + \text{Test 6} + \text{Test 7}}{\text{Number of Tests Performed during the Seven Days}} \quad (3.12)$$

Example 3.55

Problem:

The aeration tank solids concentration is determined daily. The test results for the first 10 days of the month are shown below. What is the seven-day moving average concentration on Days 7, 8, 9, and 10?

Day 1	2,330 mg/L
Day 2	3,360 mg/L
Day 3	2,640 mg/L
Day 4	2,755 mg/L
Day 5	2,860 mg/L
Day 6	2,650 mg/L
Day 7	2,340 mg/L
Day 8	2,350 mg/L
Day 9	2,888 mg/L
Day 10	2,330 mg/L

$$\text{Seven-Day Moving Average for Day 7} = \frac{2{,}330 + 3{,}360 + 2{,}640 + 2{,}755 + 2{,}860 + 2{,}650 + 2{,}340}{7}$$

$$= 2{,}705 \text{ mg/L}$$

$$\text{Seven-Day Moving Average for Day 8} = \frac{3{,}360 + 2{,}640 + 2{,}755 + 2{,}860 + 2{,}650 + 2{,}340 + 2{,}350}{7}$$

$$= 2{,}708 \text{ mg/L}$$

$$\text{Seven-Day Moving Average for Day 9} = \frac{2{,}640 + 2{,}755 + 2{,}860 + 2{,}650 + 2{,}340 + 2{,}350 + 2{,}888}{7}$$

$$= 2{,}640 \text{ mg/L}$$

$$\text{Seven-Day Moving Average for Day 10} = \frac{2{,}755 + 2{,}860 + 2{,}650 + 2{,}340 + 2{,}350 + 2{,}888 + 2{,}330}{7}$$

$$= 2{,}596 \text{ mg/L}$$

3.8 MASS BALANCE

The simplest way to express the fundamental engineering principle of *mass balance* is to say, "Everything has to go somewhere." More precisely, the *law of conservation of mass* says that when chemical reactions take place, matter is neither created nor destroyed. What this important concept allows us to do is track materials, that is, pollutants, microorganisms, chemicals, and other materials from one place to another.

The concept of mass balance plays an important role in wastewater treatment plant operations where we assume a balance exists between the material entering and leaving the treatment plant or a treatment process: "what comes in must equal what goes out." The concept is very helpful in evaluating biological systems, sampling and testing procedures, and many other unit processes within the treatment system.

In the following sections, we illustrate how the mass balance concept is used to determine the quantity of solids entering and leaving settling tanks and mass balance using BOD removal.

3.8.1 MASS BALANCE FOR SETTLING TANKS

The mass balance for the settling tank calculates the quantity of solids entering and leaving the unit.

✓ *Note:* The two numbers (in and out) must be within 10–15% of each other to be considered acceptable. Larger discrepancies may indicate sampling errors or increasing solids levels in the unit or undetected solids discharge in the tank effluent.

To get a better feel for how the mass balance for settling tanks procedure is formatted for actual use, consider the diagram below with the accompanying steps. Then, we use an example computation to demonstrate how mass balance is actually used in wastewater operations.

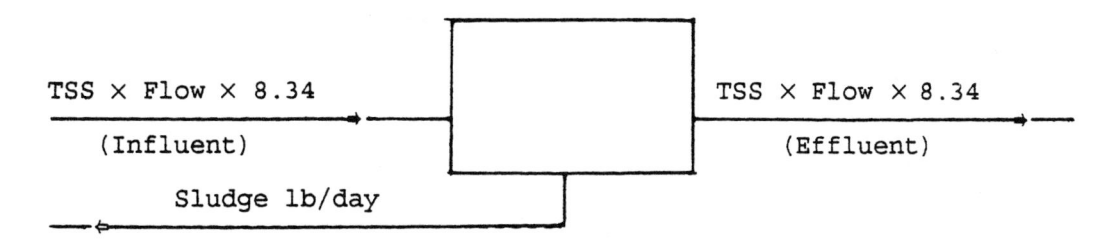

```
TSS × Flow × 8.34                    TSS × Flow × 8.34
─────────────────→      ┌─────┐      ─────────────────→
   (Influent)           │     │         (Effluent)
                        │     │
                        └──┬──┘
         Sludge lb/day     │
      ←───────────────────┘
```

Step 1: Solids in = Pounds of Influent Suspended Solids
Step 2: Pounds of Effluent Suspended Solids
Step 3: Sludge Solids Out = Pounds of Sludge Solids Pumped Per Day
Step 4: Solids In—(Solids Out + Sludge Solids Pumped)

Example 3.56

Problem:

The settling tank receives a daily flow of 4.20 MGD. The influent contains 252 mg/L suspended solids, and the unit effluent contains 140 mg/L suspended solids. The sludge pump operates 10 min/h and removes sludge at the rate of 40 gpm. The sludge is 4.2% solids. Determine if the mass balance for solids removal is within the acceptable 10–15% range.

Solution:

Step 1: Solids In = 252 mg/L × 4.20 MGD × 8.34 = 8,827 lb/day
Step 2: Solids Out = 140 mg/L × 4.20 MGD × 8.34 = 4,904 lb/day
Step 3:

$$\text{Sludge Solids} = 10\,\frac{\text{min}}{\text{hr}} \times 24\,\frac{\text{hr}}{\text{day}} \times 40\,\text{gpm} \times 8.34 \times 0.042$$

$$= 3{,}363\,\text{lb/day}$$

Step 4: Balance = 8,827 lb/day − (4,904 lb/day + 3,363 lb/day)

= 560 lb or 6.3%

3.8.2 MASS BALANCE USING BOD REMOVAL

The amount of BOD removed by a treatment process is directly related to the quantity of solids the process will generate. Because the actual amount of solids generated will vary with operational conditions and design, exact figures must be determined on a case-by-case basis. However, research has produced general conversion rates for many of the common treatment processes. These values are given in Table 3.4 and can be used if plant-specific information is unavailable.

Using these factors, the mass balance procedure determines the amount of solids the process is anticipated to produce. This is compared with the actual sludge production to determine the accuracy of the sampling and/or the potential for solids buildup in the system or unrecorded solids discharges.

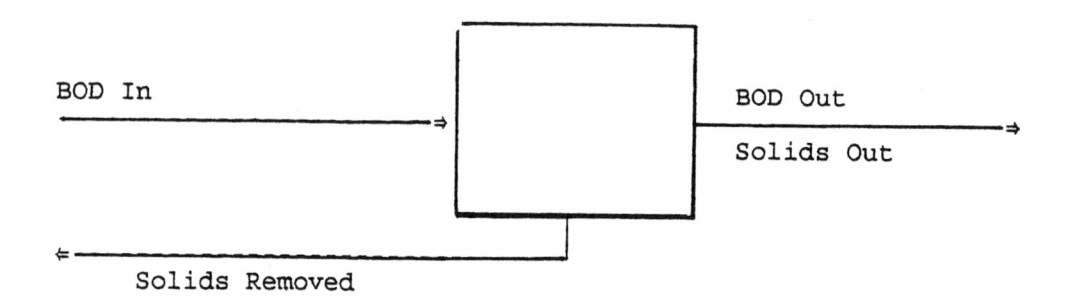

Step 1: BOD_{in} = Influent BOD × Flow × 8.34
Step 2: BOD_{out} = Effluent BOD × Flow × 8.34
Step 3: BOD Pound Removed = $BOD_{in} - BOD_{out}$
Step 4: Solids Generated, lb = BOD Removed, lb × Factor
Step 5: Solids Removed = Sludge Pumped, gpd × % Solids × 8.34
Step 6: Solids Out = Effluent Solids, mg/L × Flow, MGD × 8.34

Example 3.57

Problem:

A conventional activated sludge system with primary treatment is operating at the levels listed below. Does the mass balance for the activated sludge system indicate a problem exists?

TABLE 3.4. General Conversion Rates.

Process Type	Conversion Factor (lb solids/lb BOD Removal)
Primary Treatment	1.7
Trickling Filters	1.0
Rotating Biological Contactors	1.0
Activated Sludge w/Primary	0.7
Activated Sludge without Primary	
Conventional	0.85
Extended Air	0.65
Contact Stabilization	1.0

Plant Influent BOD	250 mg/L
Primary Effluent BOD	166 mg/L
Activated Sludge System Effluent BOD	25 mg/L
Activated Sludge System Effluent TSS	19 mg/L
Plant Flow	11.40 MGD
Waste Concentration	6,795 mg/L
Waste Flow	0.15 MGD

Solution:

$$BOD_{in} = 166 \text{ mg/L} \times 11.40 \text{ MGD} \times 8.34 = 15,783 \text{ lb/day}$$

$$BOD_{out} = 25 \text{ mg/L} \times 11.40 \text{ MGD} \times 8.34 = 2,377 \text{ lb/day}$$

$$\text{BOD Removed} = 15,783 \text{ lb/day} - 2,377 \text{ lb/day} = 13,406 \text{ lb/day}$$

$$\text{Solids Produced} = 13,406 \text{ lb/day} \times 0.7 \text{ lb Solids/lb BOD} = 9,384 \text{ lb Solids/day}$$

$$\text{Solids Removed} = 6,795 \text{ mg/L} \times 0.15 \text{ MGD} \times 8.34 = 8,501 \text{ lb/day}$$

$$\text{Difference} = 9,384 \text{ lb/day} - 8,501 \text{ lb/day} = 883 \text{ lb/day or } 9.4\%$$

This is within the acceptable range for results.

✓ We have demonstrated two ways in which mass balance can be used. However, it is important to note that the mass balance concept can be used for all aspects of wastewater and solids treatment. In each case, the calculations must take into account all of the sources of material entering the process and all of the methods available for removal of solids.

3.9 REFERENCES

Price, J. K., *Basic Math Concepts: For Water and Wastewater Operators.* Lancaster, PA: Technomic Publishing Company, Inc., 1991.

Spellman, F. R., *The Science of Air: Concepts and Applications.* Lancaster, PA: Technomic Publishing Company, Inc., 1999.

3.10 CHAPTER REVIEW QUESTIONS

Note: Answers to chapter review questions are contained in Appendix A.

3-1 $56 - 7 + 25 \div 5 \times 7 \times 3 \div 15 - 7 \times 8 =$ _____?

3-2 What is the circumference in feet of a circle that is 140 ft in diameter?

3-3 What is the area in square ft of a rectangle that is 120 ft long and 60 ft wide?

3-4 A circular tank is 20 ft deep at the outer wall. The bottom of the tank is cone-shaped, and the depth at the center of the tank is 25 ft. If the tank is 80 ft in diameter, what is the total volume of the tank including the cone-shaped portion at the bottom?

3-5 The sludge contains 6.50% solids. If 8,000 gal of sludge are removed from the primary settling tank, how many pounds of solids are removed?

3-6 The lab test indicates that a 600-g sample of sludge contains 20 g of solids. What is the percent solids in the sludge sample?

3-7 The lab test indicates that a 500-g sample of sludge contains 20 g of solids. What is the percent solids in the sludge sample?

The following information is used for review questions 3-8 to 3-12:

Primary Settling	Number	3
	Length	130 ft
	Width	110 ft
	Water Depth	12 ft
Aeration Tank	Number	4
	Length	250 ft
	Width	110 ft
	Water Depth	14 ft
Secondary Settling Tank	Number	4
	Diameter	100 ft
	Water Depth	18 ft

3-8 The effluent weir on the secondary settling tank is located along the outer edge of the tank. What is the weir length in feet for each settling tank?

3-9 What is the surface area of each of the primary settling tanks in square feet?

3-10 What is the volume of each of the aeration tanks in cubic feet?

3-11 You wish to install a fence around each aeration tank to prevent falls into the tanks. How many feet of fence must be ordered?

3-12 The secondary settling tanks consist of a cylindrical section 18 ft deep and a cone-shaped bottom that has a depth of 10 ft. What is the total volume of each settling tank in cubic feet?

The following information is provided for use in solving questions 3-13 to 3-15.

	Day 1	Day 2	Day 3	Day 4	Day 5	Day 6	Day 7	Day 8	Day 9
BOD	19	21	34	45	12	17	78	53	10
TSS	23	34	19	31	5	27	93	34	9
Fecal	1	21	123	2	10	230	13	23	1,100

3-13 What is the average BOD for the period covered by the data provided above?

3-14 What is the seven-day moving average for TSS on Day 8 and Day 9?

3-15 What is the geometric mean of the fecal results for the period covered by the data provided in the chart?

CHAPTER 4

Conversions

4.1 CONVERSION FACTORS

CONVERSION factors are used to change measurements or calculate values from one unit of measure to another.

✓ In making the conversion from one unit to another, you must know two things:

(1) The exact number that relates the two units
(2) Whether to multiply or divide by that number

For example, in converting from inches to feet, you must know that there are 12 in. in 1 ft, and you must know whether to multiply or divide the number of inches by 0.08333 (i.e., 1 in. = 0.08 ft).

When making conversions, there is often confusion about whether to multiply or divide; on the other hand, the number that relates the two units is usually known and, thus, is not a problem. Gaining understanding of the proper methodology—"mechanics"—to use for various operations requires practice.

Along with using the proper "mechanics" and much practice in making conversions, probably the easiest and fastest method of converting units is to use a conversion table.

An example of the type of conversions the wastewater operator must be familiar with is provided in the following section on temperature conversions.

4.1.1 TEMPERATURE CONVERSIONS

Most wastewater operators are familiar with the formulas used for Fahrenheit and Celsius temperature conversions:

$$°C = 5/9 \ (°F - 32) \tag{4.1}$$

$$°F = 9/5 \ (°C) + 32 \tag{4.2}$$

The difficulty arises when one is required to recall these formulas from memory.

✓ Probably the easiest way to recall these important formulas is to remember three basic steps for both Fahrenheit and Celsius conversions:

(1) Add 40°
(2) Multiply by the appropriate fraction (5/9 or 9/5)
(3) Subtract 40°

Obviously, the only variable in this method is the choice of 5/9 or 9/5 in the multiplication step. To make the proper choice, you must be familiar with two scales. On the Fahrenheit scale, the freezing point of water is 32°, whereas it is 0° on the Celsius scale. The boiling point of water is 212° on the Fahrenheit scale and 100° on the Celsius scale.

What does all this mean?

✓ Well, it is important to note, for example, that at the same temperature, higher numbers are associated with the Fahrenheit scale and lower numbers with the Celsius scale. This is an important relationship that helps you decide whether to multiply by 5/9 or 9/5. Let's look at a few conversion problems to see how the three-step process works.

Example 4.1

Suppose that you wish to convert 220°F to Celsius. Using the three-step process, we proceed as follows:

(1) Step 1: add 40°

$$220° + 40° = 260°$$

(2) Step 2: 260° must be multiplied by either 5/9 or 9/5. Because the conversion is to the Celsius scale, you will be moving to a number *smaller* than 260. Through reason and observation, it is obvious that if 260 is multiplied by 9/5, it would almost be the same as multiplying by 2, which would double 240 rather than make it smaller. On the other hand, if you multiply by 5/9, it is about the same as multiplying by 1/2, which would cut 260 in half. Because in this problem you wish to move to a smaller number, you should multiply by 5/9:

$$(5/9)(260°) = 144.4°C$$

(3) Step 3: now subtract 40°

$$144.4°C - 40.0°C = 104.4°C$$

Therefore, 220°F = 104.4°C

Example 4.2

Convert 22°C to Fahrenheit.

(1) Step 1: add 40°

$$22° + 40° = 62°$$

Because you are converting from Celsius to Fahrenheit, you are moving from a smaller to a larger number, and 9/5 should be used in the multiplication:

(2) Step 2:

$$(9/5)(62°) = 112°C$$

(3) Step 3: subtract 40°

$$112° - 40° = 72°$$

Thus, 22°C = 72°F

Obviously, it is useful to know how to make these temperature conversion calculations. However, in practical *in situ* or *non-in situ* operations, you may wish to use a temperature conversion table.

4.2 THE CONVERSION TABLE

The conversion table (Table 4.1) and Examples 4.3 through 4.20 provide many of the conversion factors used in wastewater treatment. Other conversions are presented in appropriate parts of the handbook.

✓ To convert in the opposite direction (i.e., inches to feet) divide by the factor rather than multiply.

Example 4.3

Cubic feet to gallons.

$$\text{Gallons} = \text{Cubic Feet, ft}^3 \times \text{gallons}/\text{ft}^3$$

TABLE 4.1. Conversion Table.

To Convert	Multiply By	To Get
Feet	12	Inches
Yards	3	Feet
Yards	36	Inches
Inches	2.54	Centimeters
Meters	3.3	Feet
Meters	100	Centimeters
Meters	1,000	Millimeters
Square Yards	9	Square Feet
Square Feet	144	Square Inches
Acres	43,560	Square Feet
Cubic Yards	27	Cubic Feet
Cubic Feet	1,728	Cubic Inches
Cubic Feet (Water)	7.48	Gallons
Cubic Feet (Water)	62.4	Pounds
Acre-Feet	43,560	Cubic Feet
Gallons (Water)	8.34	Pounds
Gallons (Water)	3.785	Liters
Gallons (Water)	3,785	Milliliters
Gallons (Water)	3,785	Cubic Centimeters
Gallons (Water)	3,785	Grams
Liters	1,000	Milliliters
Days	24	Hours
Days	1,440	Minutes
Days	86,400	Seconds
Million Gallons/Day	1,000,000	Gallons/Day
Million Gallons/Day	1.55	Cubic Feet/Second
Million Gallons/Day	3.069	Acre-Feet/Day
Million Gallons/Day	36.8	Acre-Inches/Day
Million Gallons/Day	3,785	Cubic Meters/Day
Gallons/Minute	1,440	Gallons/Day
Gallons/Minute	63.08	Liters/Minute
Pounds	454	Grams
Grams	1,000	Milligrams
Pressure, PSI	2.31	Head, ft (Water)
Horsepower	33,000	Foot-Pounds/Minute
Horsepower	0.746	Kilowatts
To Get	Divide By	To Convert

Sample Problem:

How many gallons of sludge can be pumped to a digester that has 3,400 ft³ of volume available?

$$\text{Gallons} = 3{,}400 \text{ ft}^3 \times 7.48 \text{ gal/ft}^3 = 25{,}432 \text{ gallons}$$

Example 4.4

Gallons to cubic feet.

$$\text{Cubic Feet} = \frac{\text{gal}}{7.48 \text{ gal/ft}^3}$$

Sample Problem:

How many cubic feet of sludge are removed when 16,200 gal are withdrawn?

$$\text{Cubic Feet} = \frac{16{,}200 \text{ gal}}{7.48 \text{ gal/ft}^3} = 2{,}166 \text{ ft}^3$$

Example 4.5

Gallons to pounds.

$$\text{Pounds, lb} = \text{Gal} \times 8.34 \text{ lb/gal}$$

Sample Problem:

If 1,450 gal of solids are removed from the primary settling tank, how many pounds of solids are removed?

$$\text{Pounds} = 1{,}450 \text{ gal} \times 8.34 \text{ lb/gal} = 12{,}093 \text{ lb}$$

Example 4.6

Pounds to gallons.

$$\text{Gallons} = \frac{\text{lb}}{8.34 \text{ lb/gal}}$$

Sample Problem:

How many gallons of water are required to fill a tank that holds 6,540 lb of water?

$$\text{Gallons} = \frac{6{,}540 \text{ lb}}{8.34 \text{ lb/gal}} = 784.2 \text{ gal}$$

Example 4.7

Milligram/liter to pounds.

✓ For plant control operations, concentrations in milligrams per liter or parts per million determined by laboratory testing must be converted to quantities in pounds, kilograms, pounds per day, or kilograms per day.

$$\text{Pounds} = \text{Concentration, mg/L} \times \text{Volume, MG} \times 8.34 \text{ lb/mg/L/MG}$$

Sample Problem:

The solids concentration in the aeration tank is 2380 mg/L. The aeration tank volume is 0.85 million gallons (MG). How many pounds of solids are in the tank?

$$\text{Pounds} = 2{,}380 \text{ mg/L} \times 0.85 \text{ MG} \times 8.34 \text{ lb/mg/L/MG} = 16{,}872 \text{ lb}$$

Example 4.8

Milligrams per liter to pounds per day.

$$\text{Pounds/Day} = \text{Concentration, mg/L} \times \text{Flow, MG} \times 8.34 \text{ lb/mg/L/MG}$$

Sample Problem:

How many pounds of solids are discharged per day when the plant effluent flow rate is 4.25 MGD and the effluent solids concentration is 24 mg/L?

$$\text{Pounds/Day} = 24 \text{ mg/L} \times 4.25 \text{ MGD} \times 8.34 \text{ lb/mg/L/MG} = 851 \text{ lb/day}$$

Example 4.9

Milligrams per liter to kilograms per day.

$$\text{kG/Day} = \text{Concentration, mg/L} \times \text{Volume, MG} \times 3.785 \text{ kG/mg/L/MG}$$

Sample Problem:

The effluent contains 25 mg/L of BOD_5. How many kilograms per day of BOD_5 are discharged when the effluent flow rate is 8.5 MGD?

$$\text{kG/Day} = 25 \text{ mg/L} \times 8.5 \text{ MGD} \times 3.785 \text{ kG/mg/L/MG} = 804.3 \text{ kG/day}$$

Example 4.10

Pounds to milligrams per liter.

$$\text{Concentration, mg/L} = \frac{\text{Quantity, lbs}}{\text{Volume, MG} \times 8.34 \text{ lb/mg/L/MG}}$$

Sample Problem:

The aeration tank contains 69,990 lb of solids. The volume of the aeration tank is 4.15 MG. What is the concentration of solids in the aeration tank in mg/L?

$$\text{Concentration, mg/L} = \frac{69{,}990 \text{ lb}}{4.15 \text{ MG} \times 8.34 \text{ lb/MG/mg/LG}} = 2{,}022 \text{ mg/L}$$

Example 4.11

Pounds per day to milligrams per liter.

$$\text{Concentration, mg/L} = \frac{\text{Quantity, lb/day}}{\text{Flow, MGD} \times 8.34 \text{ lb/mg/L/MG}}$$

Sample Problem:

The disinfection process uses 4,620 lb per day of chlorine to disinfect a flow of 24.2 MGD. What is the concentration of chlorine applied to the effluent?

$$\text{Concentration} = \frac{4,620}{24.2 \text{ MGD} \times 8.34 \text{ lb/MG/mg/L}} = 22.9 \text{ mg/L}$$

Example 4.12

Pounds to flow in million gallons per day.

$$\text{Flow} = \frac{\text{Quantity, lb/day}}{\text{Concentration, mg/L} \times 8.34 \text{ lb/mg/L/MG}}$$

Sample Problem:

You must remove 8,640 lb of solids from the activated sludge process per day. The waste activated sludge concentration is 5,699 mg/L. How many million gallons per day of waste activated sludge must be removed?

$$\text{Flow} = \frac{8,640 \text{ lb}}{5,699 \text{ mg/L} \times 8.34 \text{ lb/MG/mg/L}} = 0.18 \text{ MGD}$$

Example 4.13

Million gallons per day (MGD) to gallons per minute (gpm).

$$\text{Flow} = \frac{\text{Flow, MGD} \times 1,000,000 \text{ gal/MG}}{1,440 \text{ min/day}}$$

Sample Problem:

The current flow rate is 3.55 MGD. What is the flow rate in gallons per minute?

$$\text{Flow} = \frac{3.55 \text{ MGD} \times 1,000,000 \text{ gal/MG}}{1,440 \text{ min/day}} = 2465 \text{ gpm}$$

Example 4.14

Million gallons per day (MGD) to gallons per day (gpd).

$$\text{Flow} = \text{Flow, MGD} \times 1,000,000 \text{ gal/MG}$$

Sample Problem:

The influent meter reads 25.8 MGD. What is the current flow rate in gallons per day?

$$\text{Flow} = 25.8 \text{ MGD} \times 1{,}000{,}000 \text{ gal}/\text{MG} = 25{,}800{,}000 \text{ gpd}$$

Example 4.15

Million gallons per day (MGD) to cubic feet per second (cfs).

$$\text{Flow, cfs} = \text{Flow, MGD} \times 1.55 \text{ cfs}/\text{MGD}$$

Sample Problem:

The flow rate entering the grit channel is 2.69 MGD. What is the flow rate in cubic feet per second?

$$\text{Flow} = 2.69 \text{ MGD} \times 1.55 \text{ cfs}/\text{MGD} = 4.17 \text{ cfs}$$

Example 4.16

Gallons per minute (gpm) to million gallons per day (MGD).

$$\text{Flow, MGD} = \frac{\text{Flow, gpm} \times 1{,}440 \text{ min}/\text{day}}{1{,}000{,}000 \text{ gal}/\text{MG}}$$

Sample Problem:

The flow meter indicates that the current flow rate is 1,269 gpm. What is the flow rate in MGD?

$$\text{Flow} = \frac{1{,}269 \text{ gpm} \times 1{,}440 \text{ minutes per day}}{1{,}000{,}000 \text{ gal}/\text{MG}} = 1.82736 \text{ MGD}$$

✓ Unless a higher degree of accuracy is required, this number would be rounded off to two decimal places (1.83).

Example 4.17

Gallons per day (gpd) to million gallons per day (MGD).

$$\text{Flow, MGD} = \frac{\text{Flow, gpd}}{1{,}000{,}000 \text{ gal}/\text{MG}}$$

Sample Problem:

The totalizing flow meter indicates that 31,969,969 gal of wastewater have entered the plant in the past 24 hours. What is the flow rate in MGD?

$$\text{Flow} = \frac{31{,}969{,}969 \text{ gal}/\text{day}}{1{,}000{,}000 \text{ gal}/\text{MG}} = 31.969969 \text{ MGD}$$

✓ Unless a higher degree of accuracy is required, this number would be rounded to two decimal places (31.97).

Example 4.18

Flow in cubic feet per second (cfs) to million gallons per day (MGD).

$$\text{Flow, MGD} = \frac{\text{Flow, cfs}}{1.55\,\text{cfs/MG}}$$

Sample Problem:

The flow in a channel is determined to be 3.69 cubic feet per second (cfs). What is the flow rate in million gallons per day (MGD)?

$$\text{Flow} = \frac{3.69\,\text{cfs}}{1.55\,\text{cfs/MGD}} = 2.3806452\,\text{MGD}$$

✓ Unless a higher degree of accuracy is required, this number would be rounded off to two decimal places (2.38 MGD).

Example 4.19

Population Equivalent (PE) or Unit Loading Factor.

✓ When it is impossible to conduct a wastewater characterization study and other data are unavailable, population equivalent or unit per capita loading factors are used to estimate the total waste loadings to be treated.

If the BOD contribution of a discharger is known, the loading placed upon the wastewater treatment system in terms of equivalent number of people can be determined. The BOD contribution of a person is normally assumed to be 0.17 lb BOD/day.

$$\text{PE, people} = \frac{\text{BOD}_5\,\text{Contribution, lb/day}}{0.17\,\text{lb BOD}_5/\text{day/person}}$$

Sample Problem:

A new industry wishes to connect to the city's collection system. The industrial discharge will contain an average BOD concentration of 369 mg/L, and the average daily flow will be 52,000 gal per day. What is the population equivalent of the industrial discharge?

First, convert flow rate to million gallons per day:

$$\text{Flow} = \frac{52,000\,\text{gpd}}{1,000,000\,\text{gal/MG}} = 0.052\,\text{MGD}$$

Next, calculate the population equivalent:

$$\text{PE, people} = \frac{369\,\text{mg/L} \times 0.052\,\text{MGD} \times 8.34\,\text{lb/mg/L/MG}}{0.17\,\text{lb BOD/person/day}}$$

$$= 941\,\text{people/day}$$

Example 4.20

Specific gravity.

✓ *Specific gravity* is the ratio of the density of a substance to that of a standard material under standard conditions of temperature and pressure. The standard material for gases is air, and for liquids and solids, the standard material is water. Specific gravity can be used to calculate the weight of a gallon of liquid chemical.

$$\text{Chemical, wt/gal} = \text{Water, wt/gal} \times \text{Chemical's Specific Gravity}$$

Sample Problem:

The label states that the ferric chloride solution has a specific gravity of 1.65. What is the weight of 1 gallon of ferric chloride solution?

$$\text{Ferric Chloride} = 8.34 \text{ lb/gal} \times 1.65 = 13.8 \text{ lb/gal}$$

4.3 CHAPTER REVIEW QUESTIONS

4-1 The depth of water in the grit channel is 36 in. What is the depth in feet?

4-2 The operator withdraws 5,269 gal of solids from the digester. How many pounds of solids have been removed?

4-3 Sludge added to the digester causes a 1,996 cubic foot change in the volume of sludge in the digester. How many gallons of sludge have been added?

4-4 The plant effluent contains 35 mg/L solids. The effluent flow rate is 3.69 MGD. How many pounds per day of solids are discharged?

4-5 The plant effluent contains 26 mg/L of BOD_5. The effluent flow rate is 7.25 MGD. How many kilograms per day of BOD_5 are being discharged?

4-6 The operator wishes to remove 3,540 lb per day of solids from the activated sludge process. The waste activated sludge concentration is 3,524 mg/L. What is the required flow rate in million gallons per day?

4-7 The plant influent includes an industrial flow that contains 235 mg/L BOD. The industrial flow is 0.70 MGD. What is the population equivalent for the industrial contribution in people per day?

4-8 Determine the per capita characteristics of BOD and suspended solids (SS), if garbage grinders are installed in a community. Assume that the average per capita flow is 110 gal/d and that the typical average per capita contributions for domestic wastewater with ground kitchen wastes are BOD 0.21 lb/capita/d; SS is 0.28 lb/capita/d.

4-9 The label of hypochlorite solution states that the specific gravity of the solution is 1.1545. What is the weight of a gallon of the hypochlorite solution?

CHAPTER 5

Measuring Plant Performance

5.1 INTRODUCTION

To evaluate how well a plant or treatment unit process is operating, *performance efficiency* or *percent (%) removal* is used. The results can be compared with those listed in the plant's operation and maintenance (O & M) manual to determine if the facility is performing as expected. In this chapter, sample calculations often used to measure plant performance/efficiency are presented.

5.2 PLANT PERFORMANCE/EFFICIENCY

✓ *Note:* The calculation used for determining the performance (percent removal) for a digester is different from that used for performance (percent removal) for other processes. Care must be taken to select the right formula.

$$\% \text{ Removal} = \frac{[\text{Influent Concentration} - \text{Effluent Concentration}] \times 100}{\text{Influent Concentration}} \quad (5.1)$$

Example 5.1

The influent BOD_5 is 247 mg/L, and the plant effluent BOD is 17 mg/L. What is the percent removal?

$$\% \text{ Removal} = \frac{(247 \text{ mg/L} - 17 \text{ mg/L}) \times 100}{247 \text{ mg/L}} = 93\%$$

5.3 UNIT PROCESS PERFORMANCE/EFFICIENCY

Formula (5.1) is used again to determine unit process efficiency. The concentration entering the unit and the concentration leaving the unit (i.e., primary, secondary, etc.) are used to determine the unit performance.

$$\% \text{ Removal} = \frac{[\text{Influent Concentration} - \text{Effluent Concentration}] \times 100}{\text{Influent Concentration}} \quad (5.1)$$

Example 5.2

The primary influent BOD is 235 mg/L, and the primary effluent BOD is 169 mg/L. What is the percent removal?

$$\% \text{ Removal} = \frac{(235 \text{ mg/L} - 169 \text{ mg/L}) \times 100}{235 \text{ mg/L}} = 28\%$$

5.4 PERCENT VOLATILE MATTER REDUCTION IN SLUDGE

The calculation used to determine *percent volatile matter reduction* is more complicated because of the changes occurring during sludge digestion.

$$\% \text{ VM Reduction} = \frac{(\% \text{ VM}_{in} - \% \text{ VM}_{out}) \times 100}{[\% \text{ VM}_{in} - (\% \text{ VM}_{in} \times \% \text{ VM}_{out})]} \tag{5.2}$$

VM = Volatile Matter

Example 5.3

Using the digester data provided below, determine the % Volatile Matter Reduction for the digester.

Data:
Raw Sludge Volatile Matter   74%
Digested Sludge Volatile Matter   54%

$$\% \text{ Volatile Matter Reduction} = \frac{(0.74 - 0.54) \times 100}{[0.74 - (0.74 \times 0.54)]} = 59\%$$

5.5 CHAPTER REVIEW QUESTIONS

The following information is used for the chapter review questions.

Plant Influent	Flow	8.25 MGD
	Suspended Solids	350 mg/L
	BOD	225 mg/L
Primary Effluent	Flow	8.35 MGD
	Suspended Solids	144 mg/L
	BOD	175 mg/L
Active Sludge Effluent	Flow	8.35 MGD
(Plant Effluent)	Suspended Solids	17 mg/L
	BOD	24 mg/L
Anaerobic Digester	Solids In	6.6%
	Solids Out	13.4%
	Volatile Matter In	66.3%
	Volatile Matter Out	49.1%

5-1 What is the plant percent removal for BOD_5?

5-2 What is the plant percent removal of TSS?

5-3 What is the primary treatment percent removal of BOD_5?

5-4 What is the primary treatment percent removal of TSS?

5-5 What is the percent volatile matter reduction in the anaerobic digestion process?

CHAPTER 6

Hydraulic Detention Time

6.1 INTRODUCTION

THE term *detention time* or *hydraulic detention time* (HDT) refers to the average length of time (theoretical time) a drop of water, wastewater, or suspended particles remains in a tank or channel. It is calculated by dividing the water/wastewater in the tank by the flow rate through the tank. The units of flow rate used in the calculation are dependent on whether the detention time is to be calculated in seconds, minutes, hours, or days. Detention time is used in conjunction with various treatment processes, including sedimentation and coagulation-flocculation.

Generally, in practice, detention time is associated with the amount of time required for a tank to empty. The range of detention time varies with the process. For example, in a tank used for sedimentation, detention time is commonly measured in minutes.

The calculation methods used to determine detention time are illustrated in the following sections.

6.2 DETENTION TIME IN DAYS

$$\text{HDT, Days} = \frac{\text{Tank Volume, ft}^3 \times 7.48 \text{ gal/ft}^3}{\text{Flow, gal/day}} \quad (6.1)$$

Example 6.1

An anaerobic digester has a volume of 2,400,000 gal. What is the detention time in days when the influent flow rate is 0.07 MGD?

$$\text{D.T., Days} = \frac{2,400,000 \text{ gal}}{0.07 \text{ MGD} \times 1,000,000 \text{ gal/MG}}$$

D.T., Days = 34 days

6.3 DETENTION TIME IN HOURS

$$\text{HDT, Hours} = \frac{\text{Tank Volume, ft}^3 \times 7.48 \text{ gal/ft}^3 \times 24 \text{ hours/day}}{\text{Flow, gal/day}} \quad (6.2)$$

Example 6.2

A settling tank has a volume of 44,000 ft^3. What is the detention time in hours when the flow is 4.15 MGD?

$$\text{D.T., Hours} = \frac{44{,}000 \text{ ft}^3 \times 7.48 \text{ gal/ft}^3 \times 24 \text{ hr/day}}{4.15 \text{ MGD} \times 1{,}000{,}000 \text{ gal/MG}}$$

$$\text{D.T., Hours} = 1.9 \text{ hours}$$

6.4 DETENTION TIME IN MINUTES

$$\text{HDT, min} = \frac{\text{Tank Volume, ft}^3 \times 7.48 \text{ gal/ft}^3 \times 1{,}440 \text{ min/day}}{\text{Flow, gal/day}} \quad (6.3)$$

Example 6.3

A grit channel has a volume of 1,340 ft^3. What is the detention time in minutes, when the flow rate is 4.3 MGD?

$$\text{D.T., Minutes} = \frac{1{,}340 \text{ ft}^3 \times 7.48 \text{ gal/ft}^3 \times 1440 \text{ min/day}}{4{,}300{,}000 \text{ gal/MG}}$$

$$\text{D.T., Minutes} = 3.36 \text{ min}$$

✓ The tank volume and the flow rate must be in the same dimensions before calculating the hydraulic detention time.

6.5 CHAPTER REVIEW QUESTIONS

6-1 The influent flow rate to a primary settling tank is 1.35 MGD. The tank is 70 ft in length, 16 ft wide, and has a water depth of 10 ft. What is the detention time of the tank in hours?

The following information is used for the chapter review questions.

Plant Influent	Flow	8.40 MGD
Grit Channel	Number of Channels	2
	Channel Length	60 ft
	Channel Width	4 ft
	Water Depth	2.6 ft
Primary Settling	Number	2
	Length	160 ft
	Width	110 ft
	Water Depth	12 ft
Anaerobic Digester	Flow	19,000 gpd
	Volume	110,000 ft^3

6-2 What is the hydraulic detention time in hours for primary settling when both tanks are in service?

6-3 What is the hydraulic detention time in the grit channel in minutes when both channels are in service?

6-4 What is the hydraulic detention time of the anaerobic digester in days?

CHAPTER 7

Wastewater: Sources and Characteristics

7.1 INTRODUCTION

WASTEWATER treatment is designed to use the natural purification processes (self-purification processes of streams and rivers) to the maximum level possible. It is also designed to complete these processes in a controlled environment rather than over many miles of stream or river. Moreover, the treatment plant is also designed to remove other contaminants that are not normally subjected to the natural processes as well as treating the solids that are generated through the treatment unit steps. The typical wastewater treatment plant is designed to achieve many different purposes:

- protect public health
- protect public water supplies
- protect aquatic life
- preserve the best uses of the waters
- protect adjacent lands

Wastewater treatment is a series of steps. Each of the steps can be accomplished using one or more treatment processes or types of equipment. The major categories of treatment steps are

- Preliminary Treatment—removes materials that could damage plant equipment or would occupy treatment capacity without being treated
- Primary Treatment—removes settleable and flotable solids (may not be present in all treatment plants)
- Secondary Treatment—removes BOD_5 and dissolved and colloidal suspended organic matter by biological action; organics are converted to stable solids, carbon dioxide, and more organisms
- Advanced Waste Treatment—uses physical, chemical, and biological processes to remove additional BOD_5, solids, and nutrients (not present in all treatment plants)
- Disinfection—removes microorganisms to eliminate or reduce the possibility of disease when the flow is discharged
- Sludge Treatment—stabilizes the solids removed from the wastewater during treatment, inactivates pathogenic organisms, and/or reduces the volume of the sludge by removing water

The various treatment processes described above are discussed in detail later in the handbook.

Because wastewater operators are expected to have a well-rounded knowledge not only of treatment unit processes but also of the substance (the wastestream) they are treating, in this chapter, we describe the sources and various characteristics of the wastestream they treat.

7.2 WASTEWATER SOURCES

The principal sources of domestic wastewater in a community are the residential areas and commercial districts. Other important sources include institutional and recreational facilities and stormwater (runoff) and groundwater (infiltration). Each source produces wastewater with specific characteristics. In this chapter, wastewater sources are described in the following section (7.2.1), and the specific characteristics of wastewater are described in Section 7.3.

7.2.1 GENERATION OF WASTEWATER

Wastewater is generated by five major sources: human and animal wastes, household wastes, industrial wastes, stormwater runoff, and groundwater infiltration.

(1) *Human and animal wastes*—contains the solid and liquid discharges of humans and animals and is considered by many to be the most dangerous from a human health viewpoint. The primary health hazard is presented by the millions of bacteria, viruses, and other microorganisms (some of which may be pathogenic) present in the wastestream.

(2) *Household wastes*—are wastes, other than human and animal wastes, discharged from the home. Household wastes usually contain paper, household cleaners, detergents, trash, garbage, and other substances the homeowner discharges into the sewer system.

(3) *Industrial wastes*—includes industry-specific materials which can be discharged from industrial processes into the collection system. They typically contain chemicals, dyes, acids, alkalis, grit, detergents, and highly toxic materials.

(4) *Stormwater runoff*—many collection systems are designed to carry both the wastes of the community and stormwater runoff. In this type of system, when a storm event occurs, the wastestream can contain large amounts of sand, gravel, and other grit as well as excessive amounts of water.

(5) *Groundwater infiltration*—groundwater will enter older, improperly sealed collection systems through cracks or unsealed pipe joints. Not only can this add large amounts of water to wastewater flows but also additional grit.

7.2.2 CLASSIFICATION OF WASTEWATER

Wastewater can be classified according to the sources of flows: domestic, sanitary, industrial, combined, and stormwater.

(1) *Domestic (sewage) wastewater*—mainly contains human and animal wastes, household wastes, small amounts of groundwater infiltration, and small amounts of industrial wastes.

(2) *Sanitary wastewater*—consists of domestic wastes and significant amounts of industrial wastes. In many cases, the industrial wastes can be treated without special precautions. However, in some cases, the industrial wastes will require special precautions or a pretreatment program to ensure the wastes do not cause compliance problems for the wastewater treatment plant.

(3) *Industrial wastewater*—industrial wastes only. Often, the industry will determine that it is safer and more economical to treat its waste independent of domestic waste.

(4) *Combined wastewater*—is the combination of sanitary wastewater and stormwater runoff. All the wastewater and stormwater of the community is transported through one system to the treatment plant.

(5) *Stormwater*—a separate collection system (no sanitary waste) that carries stormwater runoff including street debris, road salt, and grit.

7.3 WASTEWATER CHARACTERISTICS

Wastewater contains many different substances that can be used to characterize it. The specific substances and amounts or concentrations of each will vary, depending on the source. Thus, it is difficult to "precisely" characterize wastewater. Instead, wastewater characterization is usually based on and applied to an "average domestic" wastewater.

✓ Keep in mind that other sources and types of wastewater can dramatically change the characteristics.

Wastewater is characterized in terms of its physical, chemical, and biological characteristics. In this section, we discuss the physical and chemical characteristics; biological characteristics are discussed in Chapter 8.

7.3.1 PHYSICAL CHARACTERISTICS

The *physical characteristics* of wastewater are based on color, odor, temperature, and flow.

(1) *Color*—Fresh wastewater is usually a light brownish-gray color. However, typical wastewater is gray and has a cloudy appearance. The color of the wastewater will change significantly if allowed to go septic (if travel time in the collection system increases). Typical septic wastewater will have a black color.

(2) *Odor*—Odors in domestic wastewater usually are caused by gases produced by the decomposition of organic matter or by other substances added to the wastewater. Fresh domestic wastewater has a musty odor. If the wastewater is allowed to go septic, this odor will change significantly—to a rotten egg odor associated with the production of hydrogen sulfide (H_2S).

(3) *Temperature*—The temperature of wastewater is commonly higher than that of the water supply because of the addition of warm water from households and industrial plants. However, significant amounts of infiltration or stormwater flow can cause major temperature fluctuations.

(4) *Flow*—The actual volume of wastewater is commonly used as a physical characterization of wastewater and is normally expressed in terms of gallons per person per day. Most treatment plants are designed using an expected flow of 100 to 200 gal per person per day. This figure may have to be revised to reflect the degree of infiltration or storm flow the plant receives. Flow rates will vary throughout the day. This variation, which can be as much as 50% to 200% of the average daily flow is known as the *diurnal flow variation*.

✓ *Diurnal*—means occurring in a day or each day; daily.

7.3.2 CHEMICAL CHARACTERISTICS

In describing the chemical characteristics of wastewater, the discussion generally includes topics such as organic matter, the measurement of organic matter, inorganic matter, and gases. For the sake of simplicity, in this handbook we specifically describe chemical characteristics in terms of alkalinity, biochemical oxygen demand (BOD), chemical oxygen demand (COD), dissolved gases, nitrogen compounds, pH, phosphorus, solids (organic, inorganic, suspended, and dissolved solids), and water.

(1) *Alkalinity*—is a measure of the wastewater's capability to neutralize acids. It is measured in terms of bicarbonate, carbonate, and hydroxide alkalinity. Alkalinity is essential to buffer (hold the neutral pH of) the wastewater during the biological treatment processes.

(2) *Biochemical oxygen demand (BOD)*—a measure of the amount of biodegradable matter in the wastewater. Normally measured by a five-day test conducted at 20°C. The BOD_5 domestic waste is normally in the range of 100 to 300 mg/L.

(3) *Chemical oxygen demand (COD)*—a measure of the amount of oxidizable matter present in the sample. The COD is normally in the range of 200 to 500 mg/L. The presence of industrial wastes can increase this significantly.

(4) *Dissolved gases*—gases that are dissolved in wastewater. The specific gases and normal concentrations are based upon the composition of the wastewater. Typical domestic wastewater contains oxygen in relatively low concentrations, carbon dioxide, and hydrogen sulfide (if septic conditions exist).

(5) *Nitrogen compounds*—the type and amount of nitrogen present will vary from the raw wastewater to the treated effluent. Nitrogen follows a cycle of oxidation and reduction. Most of the nitrogen in untreated wastewater will be in the forms of organic nitrogen and ammonia nitrogen. Laboratory tests exist for determination of both of these forms. The sum of these two forms of nitrogen is also measured and is known as *Total Kjeldahl Nitrogen* (TKN). Wastewater will normally contain between 20 and 85 mg/L of nitrogen. Organic nitrogen will normally be in the range of 8 to 35 mg/L, and ammonia nitrogen will be in the range of 12 to 50 mg/L.

(6) *pH*—a method of expressing the acid condition of the wastewater. pH is expressed on a scale of 1 to 14. For proper treatment, wastewater pH should normally be in the range of 6.5 to 9.0 (ideal—6.5 to 8.0).

(7) *Phosphorus*—essential to biological activity and must be present in at least minimum quantities or secondary treatment processes will not perform. Excessive amounts can cause stream damage and excessive algal growth. Phosphorus will normally be in the range of 6 to 20 mg/L. The removal of phosphate compounds from detergents has had a significant impact on the amounts of phosphorus in wastewater.

(8) *Solids*—most pollutants found in wastewater can be classified as solids. Wastewater treatment is generally designed to remove solids or to convert solids to a form that is more stable or can be removed. Solids can be classified by their chemical composition (organic or inorganic) or by their physical characteristics (settleable, flotable, colloidal). Concentration of total solids in wastewater is normally in the range of 350 to 1,200 mg/L.

Organic solids—consist of carbon, hydrogen, oxygen, and nitrogen and can be converted to carbon dioxide and water by ignition at 550°C. Also known as fixed solids or loss on ignition.

Inorganic content—mineral solids that are unaffected by ignition. Also known as fixed suspended solids or ash.

Suspended solids—will not pass through a glass fiber filter pad. Can be further classified as Total Suspended Solids (TSS), Volatile Suspended Solids, and/or Fixed Suspended Solids. Can also be separated into three components based on settling characteristics: settleable solids, flotable solids, and colloidal solids. Total suspended solids in wastewater is normally in the range of 100 to 350 mg/L.

Dissolved solids—will pass through a glass fiber filter pad. Can also be classified as Total Dissolved Solids (TDS), volatile dissolved solids, and fixed dissolved solids. Total dissolved solids is normally in the range of 250 to 850 mg/L.

(9) *Water*—always the major constituent of wastewater. In most cases, water makes up 99.5% to 99.9% of the wastewater. Even in the strongest wastewater, the total amount of contamination present is less than 0.5% of the total, and in average strength wastes, it is usually less than 0.1%.

7.4 TYPICAL DOMESTIC WASTEWATER CHARACTERISTICS

Table 7.1 is a summary of typical domestic wastewater characteristics.

TABLE 7.1. Typical Domestic Wastewater Characteristics.

Characteristic	Typical Characteristic
Color	Gray
Odor	Musty
Dissolved Oxygen	>1.0 mg/L
pH	6.5–9.0
TSS	100–350 mg/L
BOD_5	100–300 mg/L
COD	200–500 mg/L
Flow	100–200 gallons/person/day
Total Nitrogen	20–85 mg/L
Total Phosphorus	6–20 mg/L
Fecal Coliform	500,000–3,000,000 MPN/100 mL

7.5 CHAPTER REVIEW QUESTIONS

7-1 What is BOD_5?

7-2 Name three sources of wastewater, and give an example of the types of materials associated with each.

7-3 Define organic and inorganic.

7-4 Name the two types of solids based upon physical characteristics.

7-5 What is stormwater runoff, and how can it cause problems for the wastewater treatment plant?

7-6 Name three types of wastewater based upon the types of waste carried.

7-7 Give three reasons for treating wastewater.

CHAPTER 8

Wastewater Biology

... the wastewater specialist who has knowledge of microbiological concepts is equipped to operate wastewater treatment plant processes in a manner that will produce effluent of better quality (hopefully) than the water contained in the receiving body. (Spellman, 1997, p. 6)

8.1 INTRODUCTION

WASTEWATER treatment is essentially a combination of physical, chemical, and biological processes. With the exception of the addition of chemicals such as alum, ferric chloride, chlorine, and others, what happens in a wastewater treatment plant is basically the same as what goes on naturally in a stream (the stream self-purification process). The wastewater treatment plant simply speeds up the natural processes (under controlled conditions) by which wastewater purifies itself.

After receiving the physical aspects of treatment (i.e., screening, grit removal, and sedimentation) in preliminary and primary treatment, wastewater still contains some suspended solids and other solids that are dissolved in the water. In a natural stream, such substances are a source of food for protozoa, fungi, algae, and several varieties of bacteria. In secondary wastewater treatment, these same microscopic organisms (which are one of the main reasons for treating wastewater) are allowed to work as fast as they can to biologically convert the dissolved solids to suspended solids, which will physically settle out at the end of secondary treatment.

This chapter is designed to provide basic information about common wastewater organisms and the processes they use to purify wastewater.

8.2 WASTEWATER ORGANISMS

When it arrives at the treatment plant, wastewater influent typically contains millions of organisms. The majority of these organisms are non-pathogenic; however, several pathogenic organisms may also be present (these may include the organisms responsible for diseases such as typhoid, tetanus, hepatitis, dysentery, gastroenteritis, and others).

Many of the organisms found in wastewater are microscopic (microorganisms); they include algae, bacteria, protozoa (such as amoebas, flagellates, free-swimming ciliates, and stalked ciliates), rotifers, and viruses.

8.2.1 ALGAE

Algae are small plants that may range from microscopic (single cell) to macroscopic (visible without the aid of a microscope). Like most other plants, algae are literally solar cells that transform solar energy into other forms of biologically usable energy. The bright green algal form is the most beneficial to wastewater treatment because it uses its solar energy (obtained via photosynthesis) to produce oxygen. Other forms of algae in the wastestream may cause operational problems.

8.2.2 BACTERIA

Bacteria are microscopic, one-cell organisms that are present in human and animal body discharges.

> Of all the microorganisms present in the wastestream, bacteria are the most widely distributed, the smallest in size, the simplest in morphology (structure), the most difficult to classify, and the hardest to identify. Because of considerable diversity, it is even difficult to provide a descriptive definition of a bacterial organism. About the only generalization that can be made is that bacteria are single-celled plants, are procaryotic, are seldom photosynthetic, and reproduce by binary fission. (Spellman, 1997, p. 19)

Though (as stated above) difficult to classify, bacteria can be classified in many different ways including (1) the source of oxygen and process they use to survive (aerobic, anaerobic, anoxic, facultative); (2) their ability to cause disease (pathogenic or non-pathogenic); and/or (3) their shape and many other characteristics. From the treatment point of view, bacteria are important in wastewater treatment because they are the main workers (the "hungry proletariat") in the process to remove contaminants from the wastestream.

8.2.3 PROTOZOA

The protozoa ("first animals") are a large group of microscopic, eucaryotic organisms (a higher life form) of more than 50,000 known species that have adapted a form of cell to serve as the entire body; they are normally associated with less polluted waters. In wastewater treatment, they are a critical part of the purification process and can be used to indicate the condition of treatment processes. Protozoa normally associated with wastewater include amoeba, flagellates, free-swimming ciliates, and stalked ciliates.

Amoebas move through the wastewater by moving the liquids stored within their cell walls. Amoebas are associated with poor treatment or a young sludge (biomass) and are normally associated with treatment that produces an effluent high in BOD_5 and suspended solids.

Flagellates are protozoa having a single, long, hair-like projection (flagella) that is used to propel the organism through the wastewater and to attract food. The flagellated protozoa are normally associated with poor treatment and a young sludge (biomass). When they predominate, the plant effluent will contain large amounts of BOD_5 and suspended solids.

Free-swimming ciliates use their tiny, hair-like projections (cilia) to move themselves through the wastewater and to attract food. Free-swimming ciliates are normally associated with a moderate sludge age and effluent quality. When they predominate, the plant effluent will normally be turbid and contain a high amount of suspended solids.

Stalked ciliated protozoa attach themselves to the wastewater solids and use their cilia to attract food. The stalked ciliated protozoa are normally associated with a plant effluent that is very clear and contains low amounts of BOD_5 and suspended solids.

8.2.4 ROTIFERS

Rotifers are a higher life form normally associated with cleaner waters—in well-operating treatment plants. They are often used to indicate the performance of certain types of treatment systems.

8.2.5 VIRUSES

Viruses are extremely small and fragile microorganisms that can be either pathogenic or non-pathogenic.

8.3 BIOLOGICAL PROCESSES

The purpose of this section is to introduce the reader to the principal types of biological treatment processes that are currently being used for the treatment of wastewaters and to identify their applications. The four major types of biological treatment processes (derived from processes occurring in nature) discussed in the following are aerobic process, anaerobic process, anoxic process, and photosynthesis (used in ponds).

8.3.1 AEROBIC PROCESSES

Aerobic processes include suspended-growth, attached-growth, and combined suspended- and attached-growth processes used in unit processes such as activated sludge suspended-growth nitrification, aerated lagoons, aerobic digestion, trickling filters, rotating biological contactors, and others. In aerobic processes (as shown in Figure 8.1), organisms use free, elemental oxygen and organic matter together with nutrients (nitrogen and phosphorus) and trace metals (iron, etc.) to produce more organisms and stable dissolved and suspended solids and carbon dioxide. Aerobic processes are used primarily for carbonaceous BOD removal and nitrification.

8.3.2 ANAEROBIC PROCESSES

Anaerobic processes consist of suspended-growth, attached-growth, and combinations of the two used in anaerobic digestion, anaerobic contact processes, and anaerobic filter processes for carbonaceous BOD removal, waste stabilization, and denitrification. As shown in Figure 8.2, the anaerobic process consists of two steps, occurs completely in the absence of oxygen, and produces a useable by-product, methane gas.

In the first step of the process (see Figure 8.2), facultative microorganisms use the organic matter as food to produce more organisms, volatile (organic) acids, carbon dioxide, hydrogen sulfide and other gases, and some stable solids.

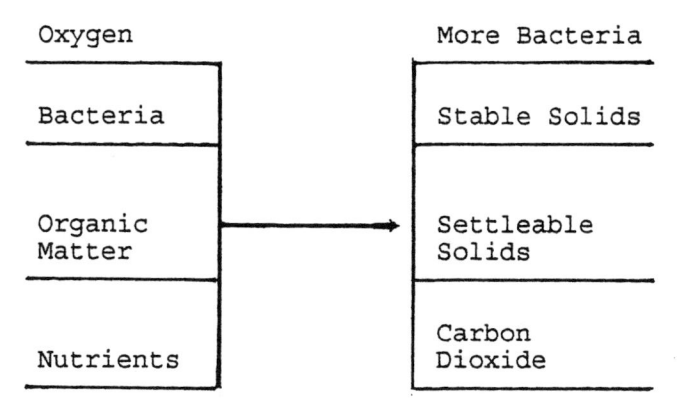

Figure 8.1 Aerobic decomposition.

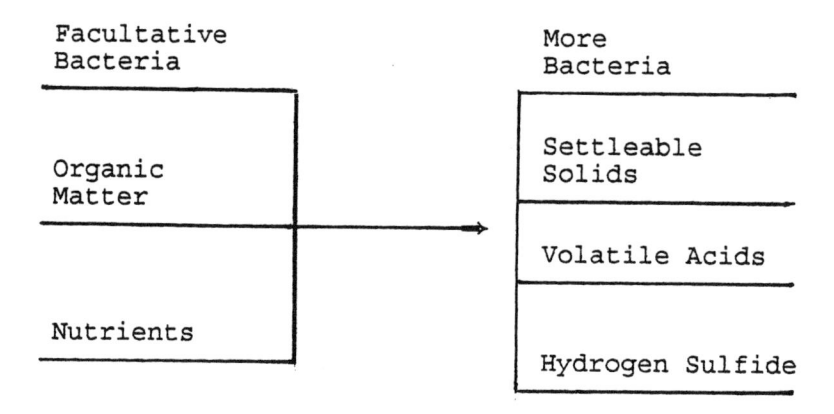

```
Facultative                    More
Bacteria                       Bacteria
┌─────────────┐                ┌──────────────────┐
│             │                │ Settleable       │
│ Organic     │                │ Solids           │
│ Matter      │                ├──────────────────┤
│             │───────────────▶│ Volatile Acids   │
│             │                ├──────────────────┤
│ Nutrients   │                │                  │
│             │                │ Hydrogen Sulfide │
└─────────────┘                └──────────────────┘

        Anaerobic Decomposition -- 1st Step


Anaerobic
Bacteria                       More Bacteria
┌─────────────┐                ┌──────────────────┐
│             │                │ Stable Solids    │
│ Volatile    │                ├──────────────────┤
│ Acids       │───────────────▶│ Settleable Solids│
│             │                ├──────────────────┤
│ Nutrients   │                │ Methane          │
└─────────────┘                └──────────────────┘

        Anaerobic Decomposition -- 2nd Step
```

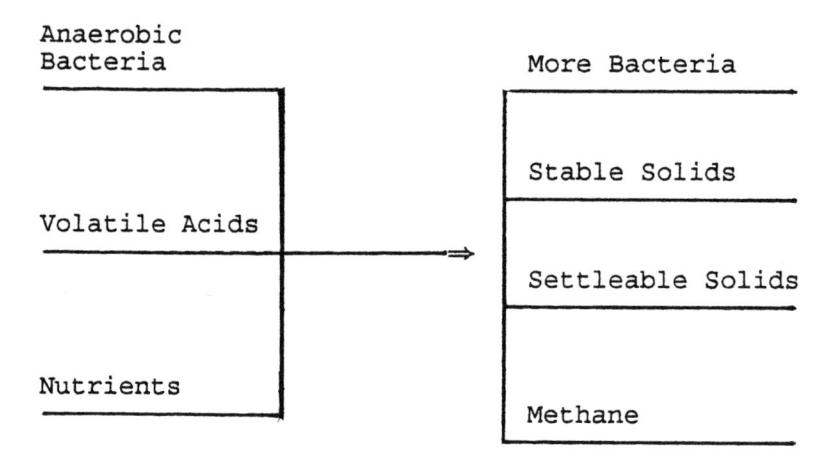

Figure 8.2 Anaerobic process.

In the second step (see Figure 8.2), anaerobic microorganisms use the volatile acids as their food source. The process produces more organisms, stable solids, and methane gas, which can be used to provide energy for various treatment system components.

8.3.3 ANOXIC PROCESSES

Anoxic processes use suspended-growth and attached-growth microorganisms in suspended-growth denitrification and fixed-film denitrification for (as you probably can guess) denitrification. As shown in Figure 8.3, in the anoxic process, microorganisms use the fixed oxygen in nitrate compounds as a source of energy. The process produces more organisms and removes nitrogen from the wastewater by converting it to nitrogen gas, which is released into the air.

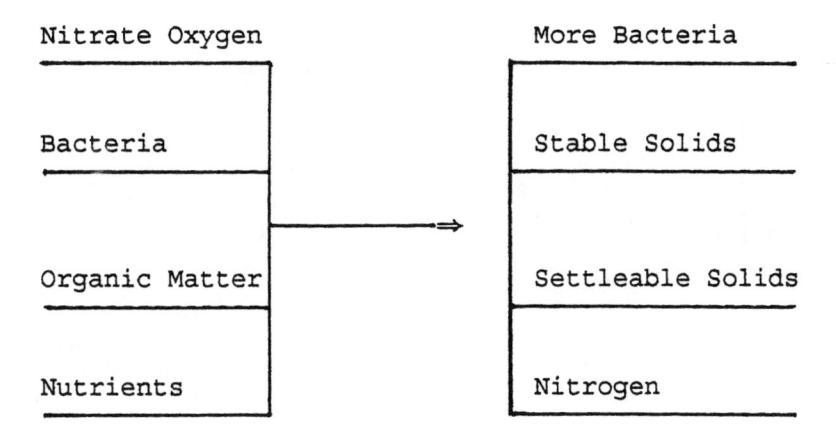

Figure 8.3 Anoxic decomposition.

8.3.4 PHOTOSYNTHESIS

Photosynthesis is used in pond processes to produce green algae that use carbon dioxide and nutrients in the presence of sunlight and chlorophyll to produce more algae and oxygen (see Figure 8.4).

8.4 THE GROWTH CURVE

The growth of bacteria can be plotted as the logarithm of cell number versus the incubation time (this information can be very useful in operating a biological treatment process). The resulting curve (see Figure 8.5) has three distinct phases. In the *Log Growth Phase*, organisms grow very rapidly, producing large numbers of new organisms. In the *Declining Growth Phase,* the rate of die-off equals the growth rate. Cell mass growth is limited by food availability. In the *Endogenous Phase,* cells must use the food accumulated in the protoplasm of the cells without replenishment.

The curve occurs when the environmental conditions required for the particular organism are reached. It is the environmental conditions (oxygen availability, pH, temperature, presence or lack

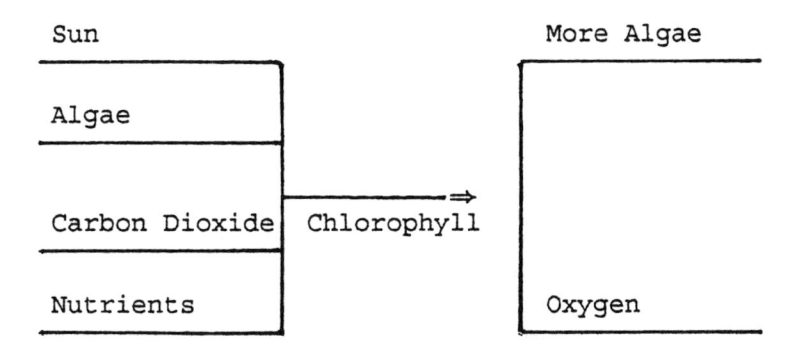

Figure 8.4 Photosynthesis.

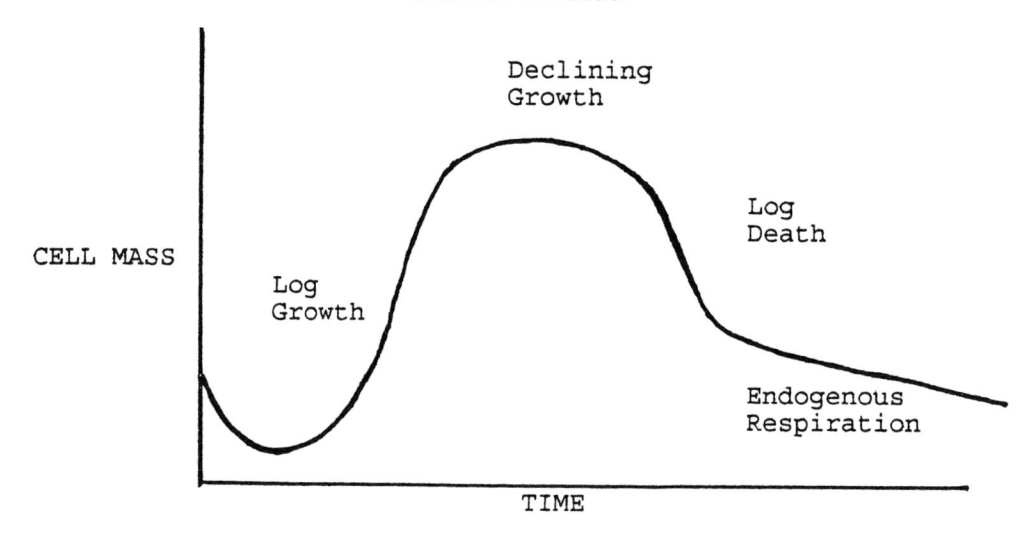

Figure 8.5 Bacterial growth curve.

of nutrients, presence or absence of toxic materials) that determine when a particular group of organisms will predominate.

✓ Note that these phases reflect the events in a population of bacteria or other microorganisms, not the individual cells. The terms *lag, declining growth,* and *death phase* do not apply to individual cells but only to populations of cells.

8.5 SELF- (NATURAL) PURIFICATION

Self- or natural purification refers to a stream or river's ability (given enough time and distance) to purify itself. For example, when wastewater is discharged to a body of moving water, natural processes occur that will remove some forms of pollution from the water (see Figure 8.6). This

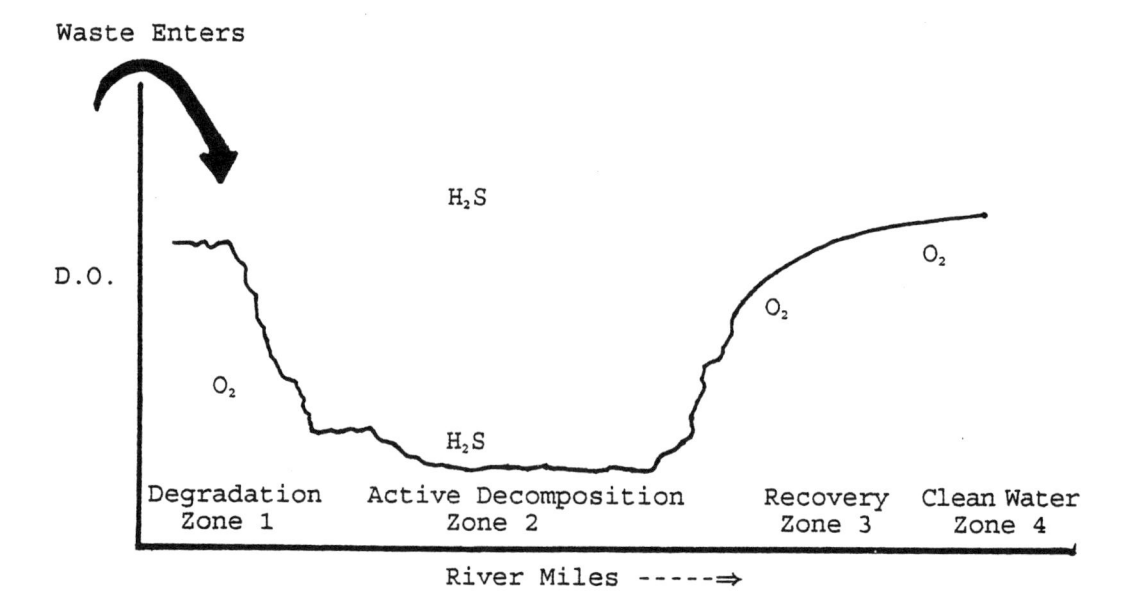

Figure 8.6 Self-purification process (in stream or river).

process has been on-going since time immemorial. It is only when the stream becomes overloaded with pollution that the natural cleaning action is retarded. When wastes were less complex than they are today, natural processes could remove the majority of pollutants. However, with the increased use of more complex chemicals and materials with high levels of toxicity and with increased population levels, the natural process has much more difficulty.

As shown in Figure 8.6, the natural process consists of four zones or stages (though it is sometimes difficult to distinguish when one zone ends and the next begins).

Let's take a closer look at each of these zones and how the self-purification process actually works.

(1) *Zone 1—Degradation*
- Wastewater enters the body of water.
- Solids begin to settle, forming sludge banks on the bottom.
- Dissolved oxygen levels in the stream decrease rapidly.
- Water takes on the characteristic color of the wastes.
- Fish population decreases rapidly.
- Bacterial population increases rapidly.

(2) *Zone 2—Active Decomposition*
- Oxygen level is zero.
- Fish life is zero.
- High concentration of bacteria and other sewage-related organisms.
- Color is black.
- Odor is rotten egg odor of hydrogen sulfide.

✓ This zone may not occur if the oxygen demand of the waste discharged does not exceed the aeration rate of the body of water.

(3) *Zone 3—Recovery*
- Oxygen level begins to increase rapidly.
- Color begins to return to normal.
- Fish population increases.
- Bacterial/microorganism population decreases.
- Odor decreases.

(4) *Zone 4—Clean Water*
- Oxygen levels at or near saturation.
- Fish populations returning to normal.
- Color, odor returning to normal.

✓ *Note:* The self-purification process removes solids that can settle and organic materials that can be removed by biological activity. It will not remove toxic materials, and it will not remove organic matter when toxic material is present until dilution reduces the concentration of the toxic material enough to eliminate the toxic effect. The process does not remove (1) disease causing organisms, (2) dissolved inorganic solids, (3) toxic materials, or (4) inorganic dyes. The process can take a long time to complete (up to 30 river miles or more), and it can be returned to degradation or decomposition by the addition of more wastes before the process is complete.

8.6 BIOGEOCHEMICAL CYCLES

Chemical elements tend to move repeatedly within the earth's crust, the ocean, and the atmosphere via living organisms. And this is a good thing—for without this constant cycling, because of the conserving nature of matter, life as we know it would not exist. This continuous and repeated movement of chemical elements or species is known as a *biogeochemical* or *nutrient cycle*.

Biogeochemical cycles are categorized into two types, the gaseous and the sedimentary. Gaseous cycles include the carbon and nitrogen cycles. The main sink of nutrients in the gaseous cycle is the atmosphere and the ocean. The sedimentary cycles include the sulfur and phosphorus cycles. The main sink for sedimentary cycles is soil and rocks of the earth's crust. In these natural cycles, the chemicals are converted from one form to another as they progress through the environment. In this handbook, it is the carbon, nitrogen, and sulfur cycles (see Figures 8.7–8.9) that require our attention because they have a major impact on the performance of the plant and may require changes in operation at various times of the year to keep them functioning properly.

8.7 REQUIREMENTS FOR BIOLOGICAL ACTIVITY

The importance of biological activity in wastewater treatment is well accepted. Maintaining optimum conditions for biological activity is dependent on maintaining the appropriate environmental conditions. The majority of wastewater processes are designed to operate using an aerobic process. The conditions required for aerobic treatment are

- sufficient oxygen (free, elemental oxygen)
- sufficient food (organic matter)
- sufficient water
- enough nutrients (nitrogen and phosphorus) to permit oxidation of the available carbon materials
- proper pH (6.5 to 9.0)
- lack of toxic materials

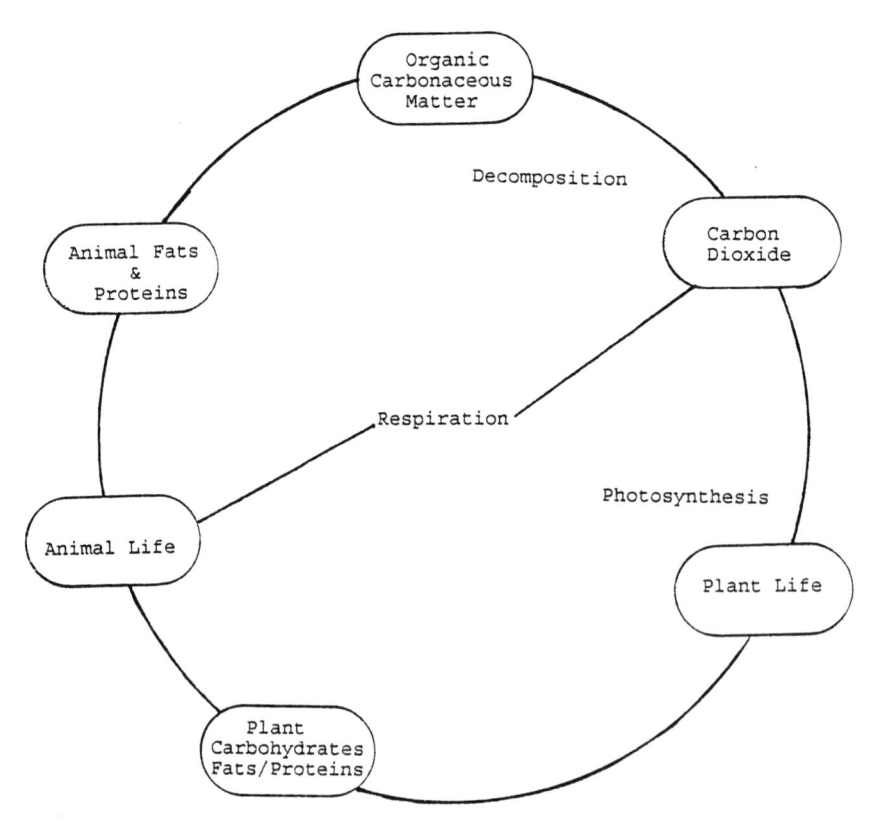

Figure 8.7 Carbon cycle (simplified).

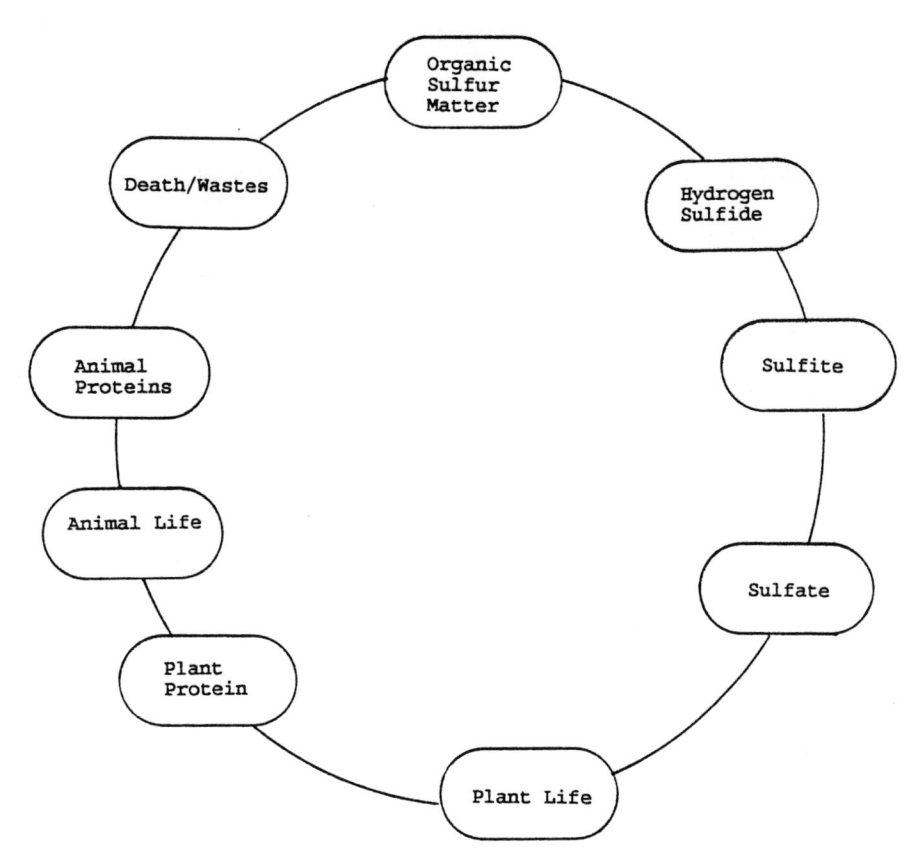

Figure 8.8 Sulfur cycle (simplified).

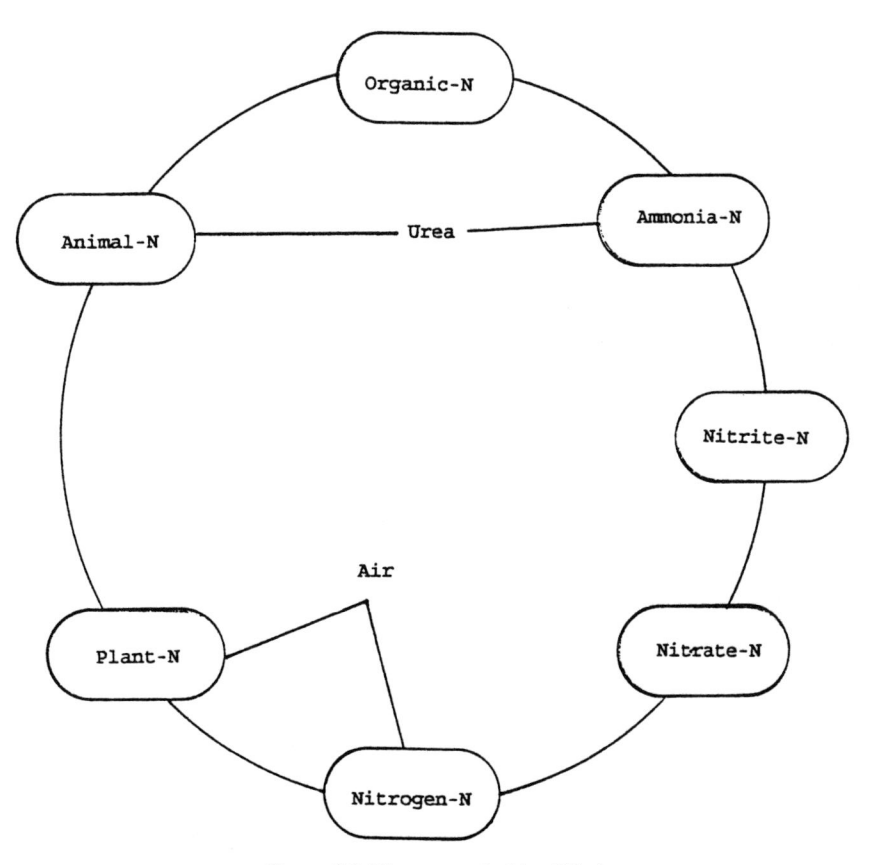

Figure 8.9 Nitrogen cycle (simplified).

83

8.8 REFERENCES

Spellman, F. R., *Microbiology for Water/Wastewater Plant Operators.* Lancaster, PA: Technomic Publishing Company, Inc., 1997.

Spellman, F. R., *The Science of Water: Concepts and Applications.* Lancaster, PA: Technomic Publishing Company, Inc., 1998.

8.9 CHAPTER REVIEW QUESTIONS

8-1 Name four types of microorganisms that may be present in wastewater.

8-2 From a human health viewpoint, which type of organism is considered the most dangerous?

8-3 What materials are produced when organic matter is decomposed aerobically?

8-4 Describe what occurs in the zone of degradation when wastes are discharged to a stream.

8-5 Which zone may disappear if the strength of the wastes discharged to the stream is reduced?

8-6 What may happen if additional wastes are discharged to a stream before the natural self-purification process is completed?

8-7 Name three materials or contaminants that are not removed by the natural self-purification process.

8-8 List three conditions that must be present for good biological activity in wastewater treatment.

CHAPTER 9

Water Hydraulics[1]

9.1 INTRODUCTION

HYDRAULICS is the study of how liquids act as they move through a channel or a pipe. Hydraulics plays an important role in the design of and operation of both the wastewater collection system and the treatment plant. In this chapter, information about several basic concepts of water hydraulics is presented.

9.1.1 TERMINOLOGY AND DEFINITIONS

- *Head*—is the equivalent distance water must be lifted to move from the supply tank or inlet to the discharge. Head can be divided into three components: static head, velocity head, and friction head.
- *Friction Head*—is the energy needed to overcome friction in the piping system. It is expressed in terms of the added system head required.
- *Static Head*—is the actual distance from the system inlet to the highest discharge point.
- *Velocity Head*—is the energy needed to keep the liquid moving at a given velocity. It is expressed in terms of the added system head required.
- *Total Dynamic Head*—is the total of the static head, friction head, and velocity head.
- *Pressure*—is the force exerted per square unit of surface area. May be expressed as pounds per square inch (psi).
- *Velocity*—is the speed of a liquid moving through a pipe, channel, or tank. May be expressed in feet per second.

9.2 BASIC CONCEPTS

$$1 \text{ ft}^3 \text{ H}_2\text{O} = 62.4 \text{ lb}$$

The relationship shown above is important: both cubic feet and pounds are used to describe a volume of water. There is a defined relationship between these two methods of measurement. The specific weight of water is defined relative to a cubic foot. One cubic foot of water weighs 62.4 lb. This relationship is true only at a temperature of 4°C and at a pressure of one atmosphere [known as standard temperature and pressure (STP)—14.7 lb per square inch at sea level]. The weight varies so little, however, that for practical purposes, we use this weight from a temperature of 0°C to 100°C.

---

[1] Much of the information contained in this chapter is adapted from F. R. Spellman, *The Science of Water: Concepts and Applications.* Lancaster, PA: Technomic Publishing Company, Inc., 1998.

✓ Remember: The important point being made here is that

$$1 \text{ ft}^3 \text{ H}_2\text{O} = 62.4 \text{ lb}$$

✓ A second relationship is also important:

$$1 \text{ gal H}_2\text{O} = 8.34 \text{ lb}$$

At standard temperature and pressure, 1 ft³ of water contains 7.48 gal. With these two relationships, we can determine the weight of 1 gal of water. This is accomplished by

$$\text{wt. of gallon of water} = 62.4 \text{ lb} \div 7.48 \text{ gal} = 8.34 \text{ lb/gal}$$

Thus,

$$1 \text{ gal H}_2\text{O} = 8.34 \text{ lb}$$

✓ Further, this information allows us to convert cubic feet to gallons by simply multiplying the number of cubic feet by 7.48 gal/ft³.

Let's take a look at how we can put this information to work.

Example 9.1

Find the number of gallons in a reservoir that has a volume of 825.5 ft³.

$$825.5 \text{ ft}^3 \times 7.48 \text{ gal/ft}^3 = 6{,}175 \text{ gal}$$

9.2.1 PROPERTIES OF WATER: TEMPERATURE/SPECIFIC WEIGHT/DENSITY

Table 9.1 is provided to show the relationship between temperature, specific weight, and density.

9.2.2 DENSITY AND SPECIFIC GRAVITY

When we say that iron is heavier than aluminum, we say that iron has greater density than aluminum. In practice, what we are really saying is that a given volume of iron is heavier than the same volume of aluminum.

✓ What is density? *Density* is the *mass per unit volume* of a substance.

Suppose you had a tub of lard and a large box of cold cereal, each having a mass of 600 g. The density of the cereal would be much less than the density of the lard because the cereal occupies a much larger volume than the lard occupies.

The density of an object can be calculated by using the formula:

$$\text{Density} = \frac{\text{Mass}}{\text{Volume}} \tag{9.1}$$

In wastewater treatment, perhaps the most common measures of density are pounds per cubic foot (lb/ft³) and pounds per gallon (lb/gal). The density of a dry material, such as cereal, lime, soda, and

TABLE 9.1.

Temperature (°F)	Specific Weight (lb/ft^3)	Density (slugs/ft^3)
32	62.4	1.94
40	62.4	1.94
50	62.4	1.94
60	62.4	1.94
70	62.3	1.94
80	62.2	1.93
90	62.1	1.93
100	62.0	1.93
110	61.9	1.92
120	61.7	1.92
130	61.5	1.91
140	61.4	1.91
150	61.2	1.90
160	61.0	1.90
170	60.8	1.89
180	60.6	1.88
190	60.4	1.88
200	60.1	1.87
210	59.8	1.86

sand, is usually expressed in pounds per cubic foot. The density of a liquid, such as liquid alum, liquid chlorine, or water, can be expressed either as pounds per cubic foot or as pounds per gallon. The density of a gas, such as chlorine gas, methane, carbon dioxide, or air, is usually expressed in pounds per cubic foot.

As shown in Table 9.1, the density of a substance like water changes slightly as the temperature of the substance changes. This happens because substances usually increase in volume (size) as they become warmer. Because of this expansion with warming, the same weight is spread over a larger volume, so the density is lower when a substance is warm than when it is cold.

✓ What is specific gravity? *Specific gravity* is the *weight of a substance compared to the weight of an equal volume of water*.

This relationship is easily seen when a cubic foot of water, which weighs 62.4 lb, is compared to a cubic foot of aluminum, which weighs 178 lb. Aluminum is 2.7 times as heavy as water.

It is not very difficult to find the specific gravity of a piece of metal. All you have to do is to weigh the metal in air, then weigh it under water. Its loss of weight is the weight of an equal volume of water. To find the specific gravity, divide the weight of the metal by its loss of weight in water.

$$\text{specific gravity} = \frac{\text{weight of substance}}{\text{weight of equal volume of water}} \qquad (9.2)$$

Let's take a look at an example problem.

Example 9.2

Suppose a piece of metal weighs 120 lb in air and 82 lb under water. What is the specific gravity?

(1) Step 1: 120 lb subtract 82 lb = 38 lb loss of weight in water

(2) Step 2:
$$\text{specific gravity} = \frac{120}{38} = 3.2$$

✓ *Note:* In a calculation of specific gravity, it is essential that the densities be expressed in the same units.

The specific gravity of water is one (1), which is the standard, the reference to which all other substances are compared. That is, any object that has a specific gravity greater than one will sink in water. Considering the total weight and volume of a ship, its specific gravity is less than one; therefore, it can float.

The most common use of specific gravity in wastewater treatment operations is in gallons-to-pounds conversions. In many cases, the liquids being handled have a specific gravity of 1.00 or very nearly 1.00 (between 0.98 and 1.08), so 1.00 may be used in the calculations without introducing significant error. However, in calculations involving a liquid with a specific gravity of less than 0.98 or greater than 1.02, the conversions from gallons to pounds must take specific gravity into account. The technique is illustrated in the following example.

Example 9.3

There are 1,255 gal of a certain liquid in a basin. If the specific gravity of the liquid is 0.93, how many pounds of liquid are in the basin?

Normally, for a conversion from gallons to pounds, we would use the factor 8.34 lb/gal (the density of water) if the substance's specific gravity was between 0.98 and 1.02. However, in this instance, the substance has a specific gravity outside this range, so the 8.34 factor must be adjusted.

Multiply 8.34 lb/gal by the specific gravity to obtain the adjusted factor:

(1) Step 1: (8.34 lb/gal) (0.93) = 7.76 lb/gal
(2) Step 2: Then, convert 1,255 gal to pounds using the corrected factor:

$$(1{,}255 \text{ gal}) (7.76 \text{ lb}/\text{gal}) = 9{,}739 \text{ lb}$$

9.2.3 FORCE AND PRESSURE

Force can be defined as the push or pull influence that causes motion. In the English system, force and weight are often used in the same way. The weight of a cubic foot of water is 62.4 lb. The force exerted on the bottom of a one foot cube is 62.4 lb. If we stack two cubes on top of one another, the force on the bottom will be 124.8 lb.

Pressure is a force per unit of area. Pounds per square inch or pounds per square foot are common expressions of pressure. The pressure on the bottom of the cube is 62.4 lb per square foot. It is normal to express pressure in pounds per square inch (psi). This is easily accomplished by determining the weight of 1 square inch of a cube 1 foot high. If we have a cube that is 12 in. on each side, the number of square inches on the bottom surface of the cube is $12 \times 12 = 144 \text{ in.}^2$. Now, by dividing the weight by the number of square inches, we can determine the weight on each square inch.

$$\text{psi} = \frac{62.4 \text{ lb}/\text{ft}}{144 \text{ in.}^2} = 0.433 \text{ psi}/\text{ft}$$

This is the weight of a column of water 1 in. square and 1 ft tall. If the column of water was 2 ft tall, the pressure would be 2 ft × 0.433 psi/ft = 0.866.

✓ Key Point: 1 ft of water = 0.433 psi

With the above information, we can convert feet of head (discussed in detail in Section 9.3) to psi by multiplying the feet of head times 0.433 psi/ft.

Example 9.4

A tank is mounted at a height of 60 ft. Find the pressure at the bottom of the tank.

Solution:

$$60 \text{ ft} \times 0.433 \text{ psi/ft} = 26.0 \text{ psi}$$

If you wanted to make the conversion of psi to feet, you would divide the psi by 0.433 psi/ft.

Example 9.5

Find the height of water in a tank if the pressure at the bottom of the tank is 18 psi.

Solution:

$$\text{height in feet} = \frac{18 \text{ psi}}{0.433 \text{ psi/ft}} = 41.6 \text{ ft}$$

9.3 HEAD

Head is the vertical distance the wastewater must be lifted from the supply tank to the discharge. The total head includes the vertical distance the liquid must be lifted (static head), the loss to friction (friction head), and the energy required to maintain the desired velocity (velocity head).

$$\text{Total Head} = \text{Static Head} + \text{Friction Head} + \text{Velocity Head} \tag{9.3}$$

9.3.1 STATIC HEAD

Static head is the actual vertical distance the liquid must be lifted.

$$\text{Static Head} = \text{Discharge Elevation} - \text{Supply Elevation} \tag{9.4}$$

Example 9.6

The supply tank is located at elevation 108 ft. The discharge point is at elevation 205 ft. What is the static head in feet?

Solution:

$$\text{Static Head, ft} = 205 \text{ ft} - 108 \text{ ft} = 97 \text{ ft}$$

9.3.2 FRICTION HEAD

Friction head is the equivalent distance of the energy that must be supplied to overcome friction.

Engineering references include tables showing the equivalent vertical distance for various sizes and types of pipes, fittings, and valves. The total friction head is the sum of the equivalent vertical distances for each component.

$$\text{Friction Head, ft} = \text{Energy Losses due to Friction} \tag{9.5}$$

9.3.3 VELOCITY HEAD

Velocity head is the equivalent distance of the energy consumed in achieving and maintaining the desired velocity in the system.

$$\text{Velocity Head, ft} = \text{Energy Losses to Maintain Velocity} \tag{9.6}$$

9.3.4 TOTAL DYNAMIC HEAD (TOTAL SYSTEM HEAD)

$$\text{Total Head} = \text{Static Head} + \text{Friction Head} + \text{Velocity Head} \tag{9.7}$$

9.3.5 PRESSURE/HEAD

The pressure exerted by wastewater is directly proportional to its depth or head in the pipe, tank, or channel. If the pressure is known, the equivalent head can be calculated.

$$\text{Head, ft} = \text{Pressure, psi} \times 2.31 \text{ ft/psi} \tag{9.8}$$

Example 9.7

The pressure gauge on the discharge line from the influent pump reads 75.3 psi. What is the equivalent head in feet?

Solution:

$$\text{Head, ft} = 75.3 \times 2.31 \text{ ft/psi} = 173.9 \text{ ft}$$

9.3.6 HEAD/PRESSURE

If the head is known, the equivalent pressure can be calculated by:

$$\text{Pressure, psi} = \frac{\text{Head, ft}}{2.31 \text{ ft/psi}} \tag{9.9}$$

Example 9.8

The tank is 15 ft deep. What is the pressure in psi at the bottom of the tank when it is filled with wastewater?

Solution:

$$\text{Pressure, psi} = \frac{15 \text{ ft}}{2.31 \text{ ft/psi}}$$

$$= 6.49 \text{ psi}$$

9.4 FLOW

The flow rate through an open channel is directly related to the velocity of the liquid and the cross-sectional area of the liquid in the channel.

$$\text{Flow }(Q), \text{cfs} = \text{Area, ft}^2 \times v, \text{fps} \tag{9.10}$$

where

Q = Flow

A = Cross-sectional area of liquid

v = velocity

Example 9.9

The channel is 5 ft wide and the water depth is 4 ft. The velocity in the channel is 4 ft per second. What is the flow rate in cubic feet per second?

Solution:

$$\text{Flow, cfs} = 5 \text{ ft} \times 4 \text{ ft} \times 4 \text{ ft/s}$$

$$= 80 \text{ cfs}$$

9.4.1 AREA/VELOCITY

At a given flow rate, the velocity of the liquid varies indirectly with changes in cross-sectional area of the channel or pipe. This principle provides the basis for many of the flow measurement devices used in open channels (weirs, flumes, and nozzles).

$$\text{Velocity}(1) \times \text{Area}(1) = \text{Velocity}(2) \times \text{Area}(2) \tag{9.11}$$

9.4.2 PRESSURE/VELOCITY

In a closed pipe flowing full (under pressure), the pressure is indirectly related to the velocity of the liquid. This principle, when combined with the principle discussed in the previous section, forms the basis for several flow measurement devices (Venturi meters and rotameters) as well as the injector used for dissolving chlorine, sulfur dioxide, and other chemicals into wastewater.

$$\text{Velocity}_1 \times \text{Area}_1 = \text{Velocity}_2 \times \text{Area}_2 \tag{9.12}$$

9.5 REFERENCE

Spellman, F. R., *The Science of Water: Concepts and Applications*. Lancaster, PA: Technomic Publishing Company, Inc., 1998.

9.6 CHAPTER REVIEW QUESTIONS

9-1 Find the number of gallons in a storage tank that has a volume of 669.9 ft^3.

9-2 Suppose a rock weighs 120 lb in air and 85 lb under water. What is the specific gravity?

9-3 There are 1,270 gal of a certain liquid in a storage tank. If the specific gravity of the liquid is 0.91, how many pounds of liquid are in the tank?

9-4 A tank is mounted at a height of 65 ft. Find the pressure at the bottom of the tank.

9-5 Find the height of water in a tank if the pressure at the bottom of the tank is 14 psi.

9-6 The elevation of the liquid in the supply tank is 2,566 ft. The elevation of the liquid surface of the discharge is 2,133 ft. What is the total static head of the system?

CHAPTER 10

Pumps

10.1 INTRODUCTION

IN wastewater collection and treatment, pumps are used to provide energy to move or lift liquids or sludges from one point to another. There are several different types of pumps that can be used for this purpose.

✓ The information provided in this chapter is for informational purposes only; it does not provide sufficient information to permit the reader to select a new or replacement pump. The specific pump required for any application must be based upon a thorough review of the application by a qualified design engineer.

10.2 TYPES OF PUMPS

According to the Hydraulic Institute (1983), all pumps may be classified as kinetic energy pumps or positive displacement pumps. The kinetic classification includes centrifugal, peripheral, and rotary type pumps. The positive displacement classification includes the screw, diaphragm, plunger, airlift, and pneumatic ejector type pumps. Each of these pump types and their principal pumping applications (in wastewater treatment) are described by Qasim (1994, pp. 178–179) and are discussed in the following.

10.2.1 KINETIC PUMPS

The centrifugal type pump consists of an impeller enclosed in a casing with inlet and discharge connections. The head is developed principally by centrifugal force. Centrifugal pumps are used to pump raw wastewater, secondary sludge return and wasting, settled primary and thickened sludge, and effluent.

Peripheral (or torque-flow/vortex) pumps consist of a recessed impeller in the side of the casing entirely out of the flow stream. A pumping vortex is set up by viscous drag. Peripheral pumps are used to pump scum, grit, sludge, and raw wastewater.

Rotary pumps consist of a fixed casing containing gears, vanes, pistons, cams, screws, etc., operating with minimum clearance. The rotating element pushes the liquid around the closed casing into the discharge pipe. Rotary pumps are normally used to pump lubricating oils, gas, chemical solutions, and small flows of water and wastewater.

10.2.2 POSITIVE DISPLACEMENT PUMPS

The screw pump uses a spiral screw operating in an inclined case. Screw pumps are used to pump grit, settled primary and secondary sludges, thickened sludge, and raw wastewater.

The diaphragm pump uses a flexible diaphragm or disk fastened over edges of a cylinder and is used primarily to pump chemicals.

Plunger pumps use a piston or plunger that operates in a cylinder. The pump discharges a definite quantity of liquid during piston or plunger movement through each stroke. Plunger pumps are normally used to pump scum, primary, secondary, and settled sludges, and chemical solutions.

The airlift pump bubbles air into a vertical tube partly submerged in water. The air bubbles reduce the unit weight of the fluid in the tube. The higher unit weight fluid displaces the low unit weight fluid, forcing it up into the tube. Airlift pumps are used in secondary sludge circulation and wasting and to pump grit.

In pneumatic ejector pumps, air is forced into the receiving chamber, which ejects the wastewater from the receiving chamber. Pneumatic ejector pumps are used to pump raw wastewater at small installations (100 to 600 L/min).

10.3 CHARACTERISTICS: CENTRIFUGAL AND POSITIVE DISPLACEMENT PUMPS

As stated above, there are two major categories of pumps: centrifugal and positive displacement pumps. Because the wastewater operator works with and/or around a wide variety of pump types that are either centrifugal or positive displacement types, and because some knowledge of pumps and their basic operations is required in most state licensing examinations, it is important for the operator to have at least some basic knowledge about pumps. Thus, in this section, we discuss the basics related to centrifugal and positive displacement pumps.

10.3.1 CENTRIFUGAL PUMPS: COMPONENTS AND CHARACTERISTICS

Centrifugal pumps are composed of an inlet and discharge, rotating shaft, either an open or semi-open impeller, volute (casing), packing assembly or mechanical seal, and a shaft lubrication system using either water, oil, or grease as the lubricant.

The capacity of centrifugal pumps is related to the total dynamic head on the system. Capacity will be zero if the total head is too great (known as cutoff head). Capacity can be changed by (1) changing the size of the impeller or (2) by changing the rotational speed of the impeller. Centrifugal pumps can be operated for short periods against a closed discharge valve but must be primed (filled with liquid) in order to operate.

✓ The most common centrifugal pump used in wastewater treatment uses a closed impeller.

10.3.2 POSITIVE DISPLACEMENT PUMPS: COMPONENTS AND CHARACTERISTICS

Positive displacement pumps use physical force to "push" material from one point to another. They are normally used for transfer of heavier materials or for precise control of the volume of liquid being pumped. The components used in positive displacement pumps depend on the type of pump.

Positive displacement pumps apply a positive pressure to the material being pumped and continue to deliver at the same rate until cutoff head is reached. Operation against a closed discharge can cause mechanical failure. They are normally used for pumping sludges or other heavy materials.

✓ Small positive displacement pumps are the pumps of choice when there is a need to deliver precise chemical doses.

10.4 OPERATING PUMPS

The wastewater operator is required to operate process pumps effectively, efficiently, and safely.

In order to do this, the operator needs to have some knowledge of packing and seals and shaft lubrication.

10.4.1 PACKING AND SEALS

To operate correctly and effectively, pumps require that the casing (volute/cylinder, etc.) be tightly sealed. The seal may be a mechanical or conventional packing seal and is designed to prevent air from entering the inside of the pump.

✓ Seal failure or improperly installed seals will reduce the pump's efficiency.

10.4.1.1 Mechanical Seals

Mechanical seals consist of one or more sets of rings (one stationary, one rotating) that prevent air from entering the pump and liquid from leaking out (see Figure 10.1).

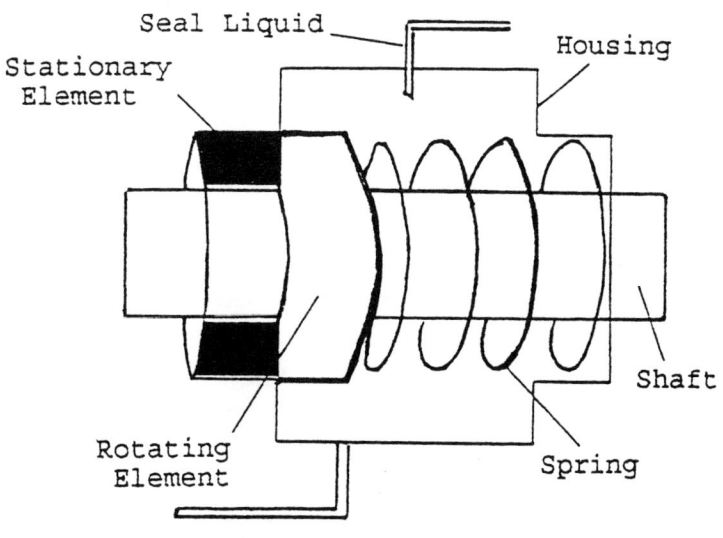

✓ Mechanical seals are normally designed to prevent leakage.

10.4.1.2 Conventional Seals

Conventional seals consist of several rings of packing material (braided teflon, hemp, and other materials), placed around the shaft and compressed with a packing gland (see Figure 10.2). This type of seal also prevents air from entering the pump and may include one or more lantern rings to distribute water or other lubricant along the shaft.

✓ Packing system is normally designed to permit 20 drops per minute leakage.

10.4.2 SHAFT LUBRICATION

Shaft lubrication is essential to prevent excessive wear and to prevent air from entering the pump casing. Lubrication may be provided by water (most common), oil, or grease.

✓ When water is used, care must be taken to avoid contamination of the water supply.

Figure 10.1 Mechanical seal.

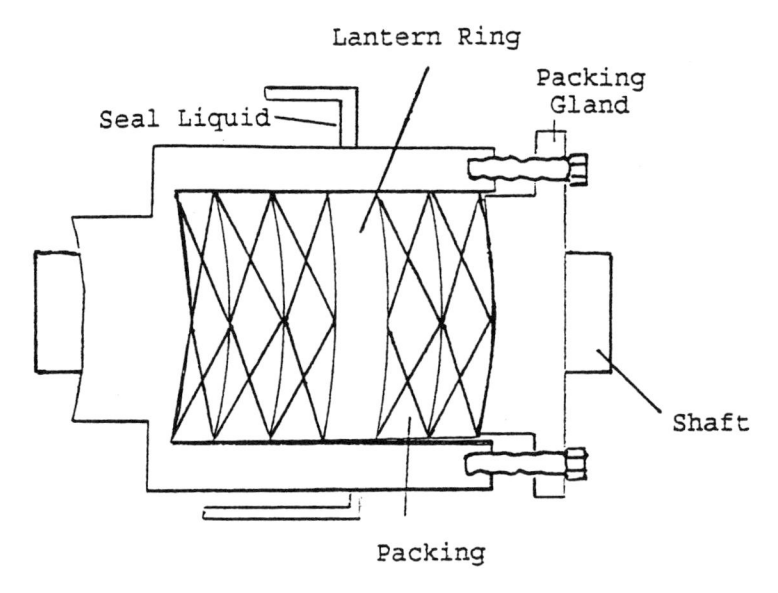

Figure 10.2 Conventional seal.

10.5 PUMP CALCULATIONS

Though most pump information and nomenclature are provided as a part of the motor-pump nameplate data, on occasion wastewater operators are called upon to make various calculations related to pumps. In this section, we discuss some of these calculations: specifically, for work, power, horsepower, water horsepower, brake horsepower, motor horsepower, and electrical power.

10.5.1 WORK

Work is defined as moving an object (weight) a vertical distance. It is expressed as foot-pounds.

$$\text{Work, ft-lb} = \text{Weight, lb} \times \text{Distance}_{\text{vertical}}\text{ ft} \tag{10.1}$$

Example 10.1

How much work is performed if 400 gal of wastewater are lifted (pumped) a vertical distance of 60 ft?

Solution:

$$\text{Work} = 400 \text{ gal} \times 8.34 \text{ lb/gal} \times 60 \text{ ft}$$

$$= 200{,}160 \text{ ft-lb}$$

10.5.2 POWER

Power is the amount of work done per unit time. It is expressed in foot-pounds per minute.

$$\text{Power} = \frac{\text{Work, ft-lb}}{\text{Elapsed Time, min}} \tag{10.2}$$

Example 10.2

How much power in foot-pounds per minute will be needed to lift 600 gal of wastewater a vertical distance of 60 ft in 4 min?

Solution:

$$\text{Power, ft-lb/min} = \frac{600 \text{ gal} \times 8.34 \text{ lb/gal} \times 60 \text{ ft}}{4 \text{ min}}$$

$$= 75,060 \text{ ft-lb/min}$$

10.5.3 HORSEPOWER

Horsepower (hp) is a common expression for power in wastewater treatment. One horsepower is equal to 33,000 ft-lb of work per minute.

$$\text{Horsepower, hp} = \frac{\text{Power, ft-lb/min}}{33,000 \text{ ft-lb/min/hp}} \qquad (10.3)$$

Example 10.3

The pump lifts 669 gpm of water a vertical distance of (head) 150 ft. How much horsepower is used?

Solution:

$$\text{Horsepower} = \frac{669 \text{ gpm} \times 8.34 \text{ lb/gal} \times 150 \text{ ft}}{33,000 \text{ ft-lb/min/hp}} = 25.4 \text{ hp}$$

10.5.4 WATER HORSEPOWER

Water horsepower (Whp) is the amount of power required to move a given volume of water a specified total head.

$$\text{Water hp} = \frac{\text{Pump Rate, gpm} \times \text{Total Dynamic Head, ft} \times 8.34 \text{ lb/gal}}{33,000 \text{ ft-lb/min/hp}} \qquad (10.4)$$

✓ Some references combine the 8.34 lb/gal and the 33,000 ft-lb/min figures into one constant. When this is done, the calculation becomes:

$$\text{Water hp} = \frac{\text{Pump Rate, gpm} \times \text{Total Dynamic Head, ft}}{33,000 \text{ ft-lb/min/hp}} \qquad (10.5)$$

Example 10.4

A pump must deliver 275 gpm to a total head of 75 feet. What is the required water horsepower?

Solution:

$$\text{Water hp} = \frac{275 \text{ gpm} \times 75 \text{ ft} \times 8.34 \text{ lb/gal}}{33,000 \text{ ft-lb/min/hp}} = 5.2 \text{ Whp}$$

10.5.5 BRAKE HORSEPOWER

Brake horsepower (Bhp) is the amount of horsepower to be supplied to the pump to provide the required water horsepower.

✓ The efficiency of the pump must be taken into account.

$$\text{Brake Horsepower, hp} = \frac{\text{Water Horsepower, hp}}{\% \text{ Efficiency}_{pump}} \quad (10.6)$$

Example 10.5

Under the specified conditions, the pump efficiency is 75%. If the required water horsepower is 5.5 hp, what is the required brake horsepower?

Solution:

$$\text{Bhp} = \frac{5.5 \text{ Whp}}{0.75} = 7.3 \text{ Bhp}$$

10.5.6 MOTOR HORSEPOWER

Motor horsepower (Mhp) is the rated horsepower of the motor necessary to produce the desired brake and water horsepower.

$$\text{Motor Horsepower, hp} = \frac{\text{Brake Horsepower, hp}}{\% \text{ Efficiency}_{motor}} \quad (10.7)$$

Example 10.6

The motor is 93% efficient. What is the required motor horsepower when the required brake horsepower is 8.0 Bhp?

Solution:

$$\text{Mhp} = \frac{8.0 \text{ Bhp}}{0.93} = 8.6 \text{ Mhp}$$

10.5.7 ELECTRICAL POWER

To determine the electrical energy required/consumed during a given period of time, the horsepower is converted to electrical energy (kilowatts), then multiplied by the hours of operation to obtain kilowatt-hours.

$$\text{Kilowatt-hour} = \text{Horsepower} \times 0.746 \text{ kW/hp} \times \text{Hours Operated} \quad (10.8)$$

Example 10.7

A 60-horsepower motor operates at full load 10 hours per day (seven days per week). How many kilowatt-hours of energy does it consume per day?

Solution:

$$\text{Kilowatt-hour/day} = 60 \text{ hp} \times 0.746 \text{ kW/hp} \times 10 \text{ hours/day}$$

$$= 448 \text{ kW-hr/day}$$

10.6 REFERENCES

Hydraulic Institute Standards for Centrifugal, Rotary and Reciprocation Pumps, 14th ed., Cleveland, OH, 1983.

Qasim, S. R., *Wastewater Treatment Plants: Planning, Design, and Operation.* Lancaster, PA: Technomic Publishing Company, Inc., 1994.

10.7 CHAPTER REVIEW QUESTIONS

Inlet Elevation	1,810 ft
Discharge Elevation	2,225 ft
Friction Head	310 ft
Velocity Head	175 ft
Pump Rate	425 gpm
Pump Efficiency	71%
Motor Efficiency	96%
Operating Time	12 hours/day, 5 days/week
Electrical Cost	$0.02333/kilowatt-hour

10-1 What is the required motor horsepower?

10-2 What is the power requirement of the system in kilowatts?

10-3 Assuming the required motor horsepower is rounded off to the next highest 25-horsepower increment (i.e., 25, 75, 100, 125 etc.), what is the cost of operating the pump for 1 year (52 weeks)?

CHAPTER 11

Wastewater Collection Systems

11.1 WASTEWATER COLLECTION SYSTEMS

As shown in the simplified representation in Figure 11.1, wastewater collection systems collect and carry wastewater to the treatment plant. The complexity of the system depends on the size of the community and the type of system selected. Methods of collection and conveyance of wastewater include gravity systems, force main systems, vacuum systems, and combinations of all three types of systems.

11.1.1 GRAVITY COLLECTION SYSTEM

In a gravity collection system, the collection lines are sloped to permit the flow to move through the system with as little pumping as possible (see Figure 11.2). The slope of the lines must keep the wastewater moving at a velocity (speed) of 2–4 ft per second. Otherwise, at lower velocities, solids will settle out causing clogged lines, overflows, and offensive odors. To keep collection systems lines at a reasonable depth, wastewater must be lifted (pumped) periodically so that it can continue flowing "downhill" to the treatment plant. Pump stations are installed at selected points within the system for this purpose (see Figure 11.2).

11.1.2 FORCE MAIN COLLECTION SYSTEM

In the force main collection system shown in Figure 11.3, wastewater is collected to central points and pumped under pressure to the treatment plant. The system is normally used for conveying wastewater long distances. The use of the force main system allows the wastewater to flow to the treatment plant at the desired velocity without using sloped lines. It should be noted that the pump station discharge lines in a gravity system are considered to be force mains because the content of the lines is under pressure.

✓ Extra care must be taken when performing maintenance on force main systems because the content of the collection system is under pressure.

11.1.3 VACUUM SYSTEM

In a vacuum collection system, wastewaters are collected to central points and then drawn toward the treatment plant under vacuum. The system consists of a large amount of mechanical equipment and requires a large amount of maintenance to perform properly. Generally, the vacuum-type collection system is not economically feasible.

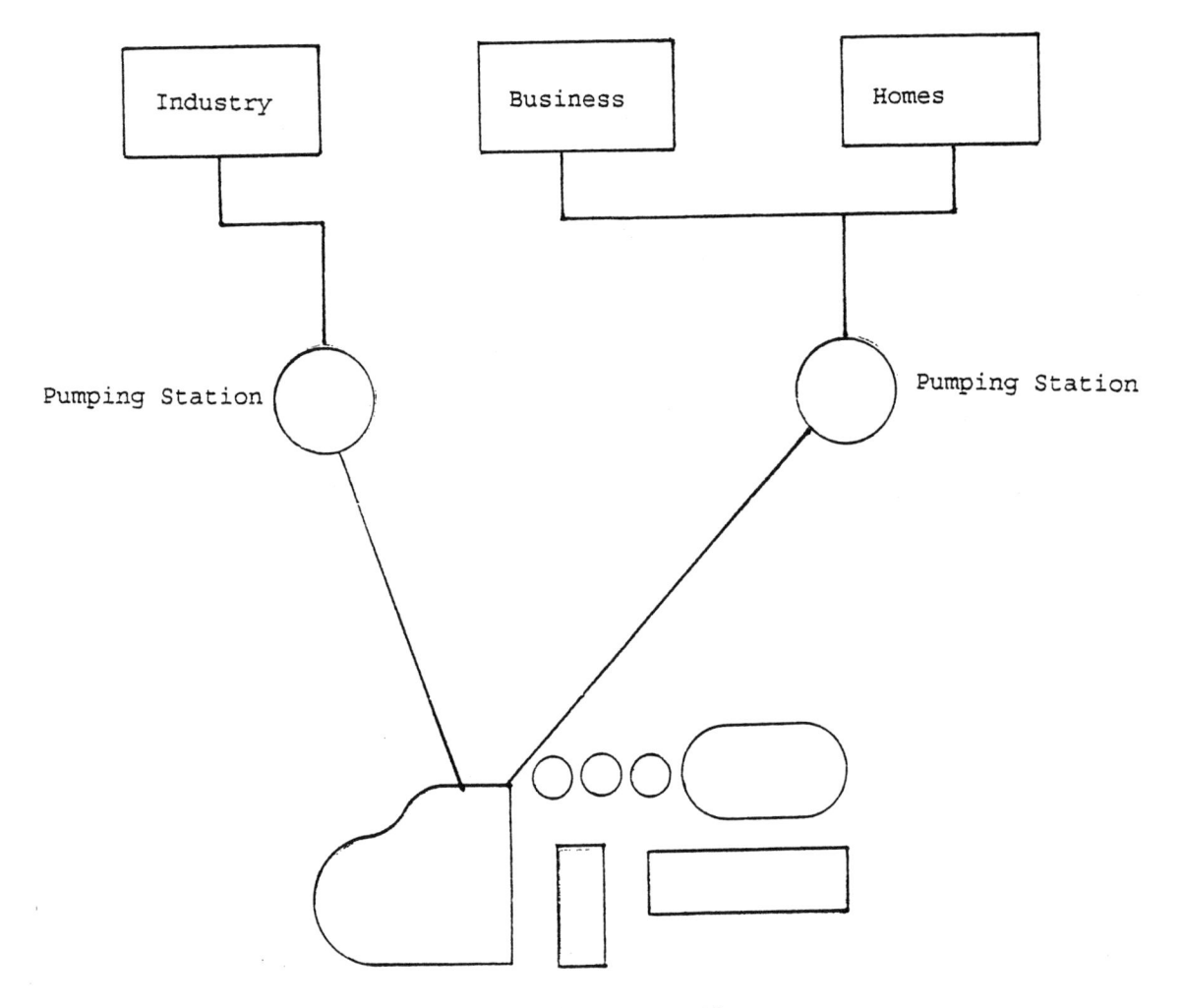

Figure 11.1 Wastewater collection system.

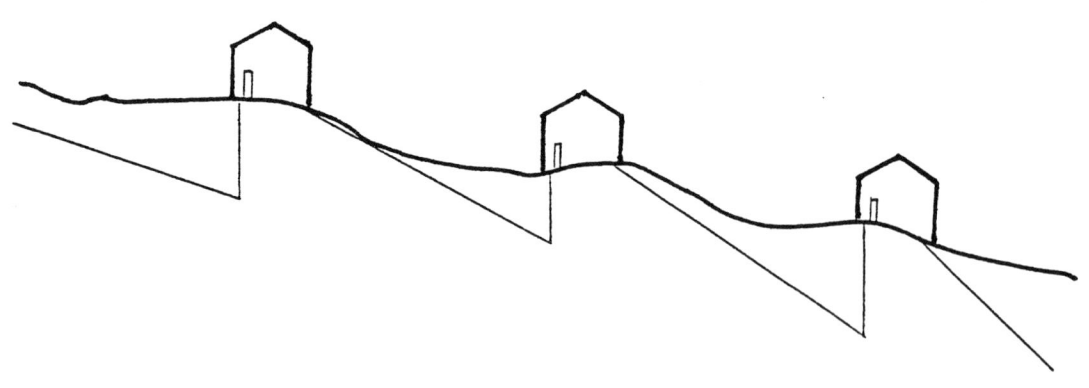

Figure 11.2 Gravity collection system.

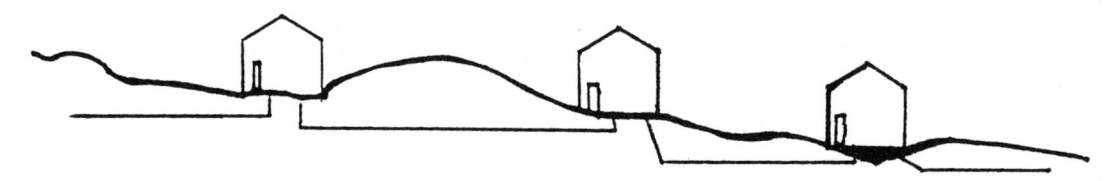

Figure 11.3 Force main collection system.

11.2 PUMPING STATIONS

Pumping stations provide the motive force (energy) to keep the wastewater moving at the desired velocity. They are used in both the force main and gravity systems. They are designed in several different configurations and may use different sources of energy to move the wastewater (i.e., pumps, air pressure, or vacuum). One of the more commonly used types of pumping station designs is the wet well/dry well design.

11.2.1 WET WELL/DRY WELL PUMPING STATIONS

The wet well/dry well pumping station consists of two separate spaces or sections separated by a common wall (see Figure 11.4). Wastewater is collected in one section (wet well section), and the pumping equipment (and, in some cases, the motors and controllers) are located in a second section known as the dry well. There are many different designs for this type of system but, in most cases, the pumps selected for this system are of a centrifugal design. There are a couple of major considerations in selecting this design: (1) it allows for the separation of mechanical equipment (pumps, motors, controllers, wiring, etc.) from the potentially corrosive atmosphere (sulfides) of the wastewater; and (2) this type of design is usually safer for workers because they can monitor, maintain, operate, and repair equipment without entering the pumping station wet well.

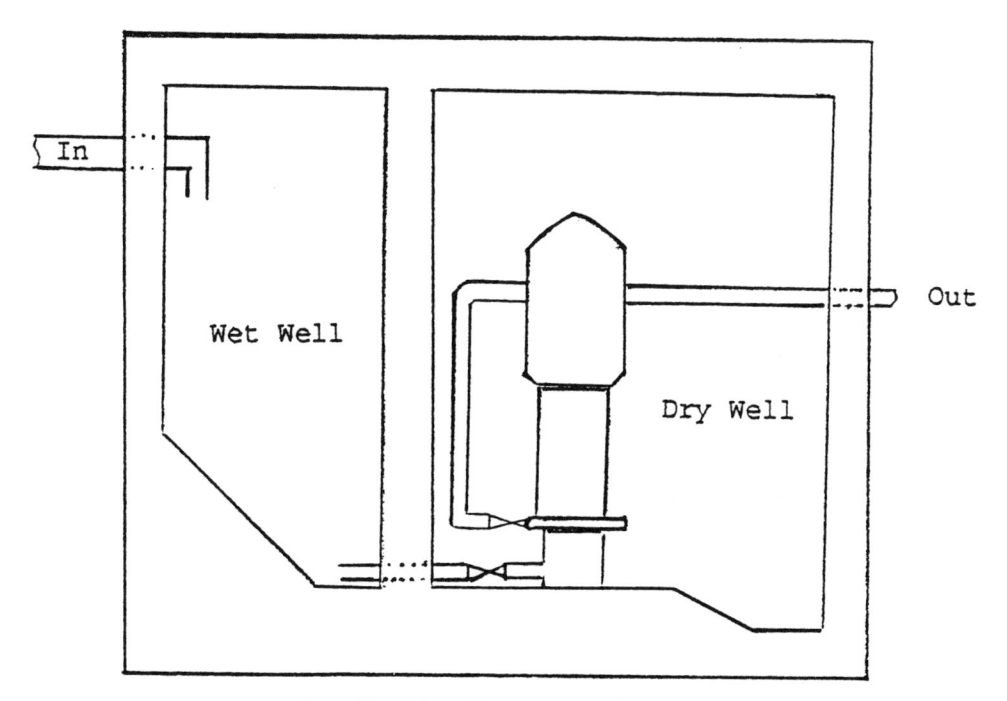

Figure 11.4 Wet well/dry well pumping station.

✓ *Note:* Most pumping station wet wells are confined spaces. To ensure safe entry into such spaces, compliance with OSHA's 29 CFR 1910.146 (Confined Space Entry Standard) is recommended.

11.2.2 WET WELL PUMPING STATIONS

Another type of pumping station design is the wet well type. This type consists of a single compartment that collects the wastewater flow [see Figure 11.5(a)]. The pump is submerged in the wastewater with motor controls located in the space or has a weather-proof motor-housing located above the wet well [see Figure 11.5(b)]. In this type of station, a submersible centrifugal pump is normally used.

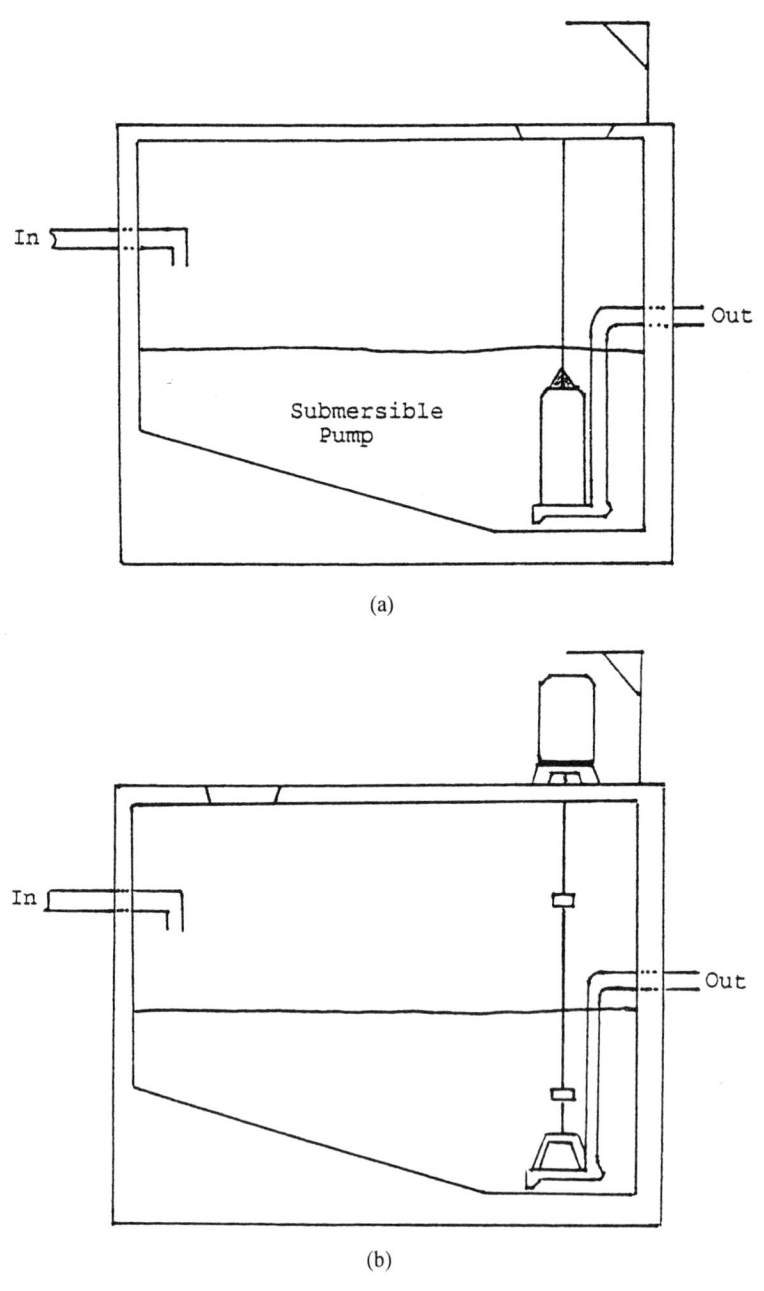

Figure 11.5 Wet well pumping stations: (a) wet well with submersible pump; (b) wet well with external motor.

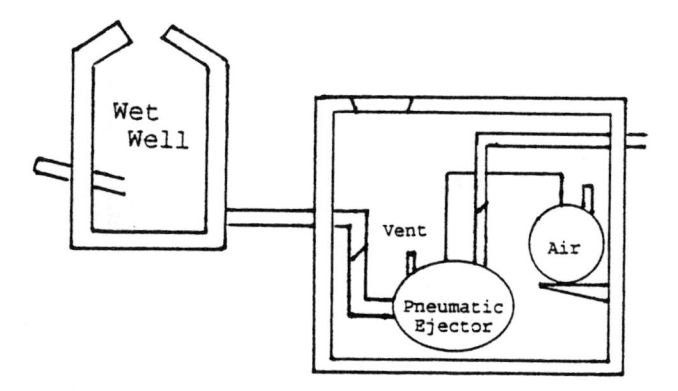

Figure 11.6 Pneumatic pumping station.

11.2.3 PNEUMATIC PUMPING STATIONS

The pneumatic pumping station consists of a wet well and a control system that controls the inlet and outlet value operations and provides pressurized air to force or "push" the wastewater through the system (see Figure 11.6). The exact method of operation depends on the system design. When operating, wastewater in the wet well reaches a predetermined level and activates an automatic valve, which closes the influent line. The tank (wet well) is then pressurized to a predetermined level. When the pressure reaches the predetermined level, the effluent line valve is opened, and the pressure pushes the wastestream out the discharge line.

11.3 PUMPING STATION WET WELL CALCULATIONS

Calculations normally associated with pumping station wet well design (determining design lift or pumping capacity, etc.) are usually left up to the design and mechanical engineers. However, on occasion, the wastewater operator or interceptor's technician may be called upon to make certain basic calculations. Usually, these calculations deal with determining either pump capacity without influent (e.g., to check the pumping rate of the station's constant speed pump) or pump capacity with influent (e.g., to check how many gallons per minute the pump is discharging). In this section, we use examples to describe instances of how and where these two calculations are made.

Example 11.1

Determining pump capacity without influent.

Problem:

A pumping station wet well is 10 ft by 9 ft. The operator needs to check the pumping rate of the station's constant speed pump. To do this, the influent valve to the wet well is closed for a 5-min test, and the level in the well dropped 2.2 ft. What is the pumping rate in gallons per minute? (See Figure 11.7.)

Solution:

Using the length and width of the well, we can find the area of the water surface.

$$10 \text{ ft} \times 9 \text{ ft} = 90 \text{ ft}^2$$

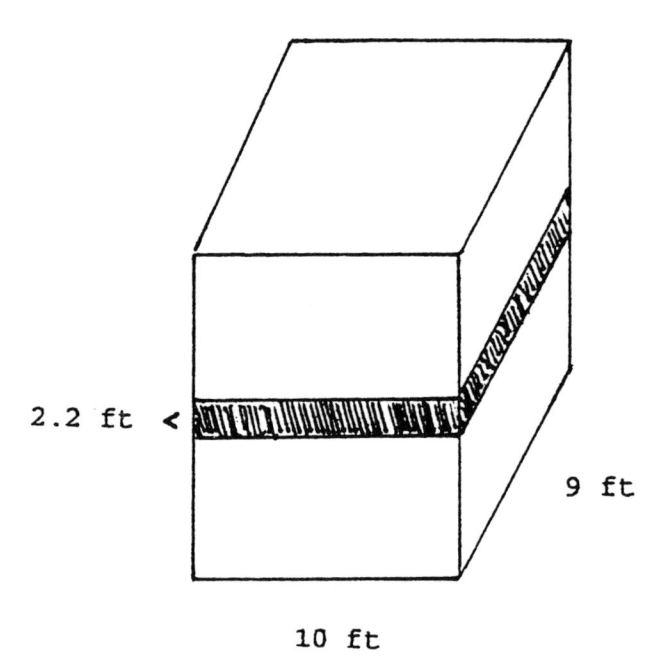

Figure 11.7 Wet well dimensions—Example 11.1.

The water level dropped 2.2 ft. From this, we can find the volume of water removed by the pump during the test.

$$\text{Area} \times \text{Depth} = \text{Volume} \tag{11.1}$$

$$90 \text{ ft}^2 \times 2.2 \text{ ft} = 198 \text{ ft}^3$$

One cubic foot of water holds 7.48 gal. We can convert this volume in cubic feet to gallons.

$$198 \text{ ft}^3 \times \frac{7.48 \text{ gal}}{1 \text{ ft}^3} = 1481 \text{ gal}$$

The test was done for 5 minutes. From this information, a pumping rate can be calculated.

$$\frac{1481 \text{ gal}}{5 \text{ min}} = \frac{296.2}{1 \text{ min}} = 296.2 \text{ gpm}$$

Example 11.2

Determining pump capacity with influent.

Problem:

A wet well is 8.2 ft by 9.6 ft. The influent flow to the well, measured upstream, is 365 gpm. If the wet well rises 2.2 in. in 5 min, how many gallons per minute is the pump discharging?

Solution:

$$\text{Influent} = \text{Discharge} + \text{Accumulation} \tag{11.2}$$

$$\frac{365 \text{ gal}}{1 \text{ min}} = \text{Discharge} + \text{Accumulation}$$

We want to calculate the discharge. Influent is known, and we have enough information to calculate the accumulation.

$$\text{Volume accumulated} = 8.2 \text{ ft} \times 9.6 \text{ ft} \times 2.2 \text{ in.} \times \frac{1 \text{ ft}}{12 \text{ in.}} \times \frac{7.48 \text{ gal}}{1 \text{ ft}^3}$$

$$= 108 \text{ gal}$$

$$\text{Accumulation} = \frac{108 \text{ gal}}{5 \text{ min}} = 21.6 \frac{\text{gal}}{\text{min}} = 21.6 \text{ gpm}$$

Using Equation (11.2):

$$\text{Influent} = \text{Discharge} + \text{Accumulation}$$

$$365 \text{ gpm} = \text{Discharge} + 21.6$$

Subtracting from both sides:

$$365 \text{ gpm} - 21.6 \text{ gpm} = \text{Discharge} + 21.6 \text{ gpm} - 21.6 \text{ gpm}$$

$$343.4 \text{ gpm} = \text{Discharge}$$

The wet well pump is discharging 343.4 gal each minute.

11.4 CHAPTER REVIEW QUESTION

11-1 Why are pumping stations required in a gravity collection system?

CHAPTER 12

Preliminary Treatment

12.1 INTRODUCTION

THE initial stage of treatment in the wastewater treatment process (following collection and influent pumping) is preliminary treatment. Raw influent entering the treatment plant may contain many kinds of materials (trash). The purpose of preliminary treatment is to protect plant equipment by removing these materials, which could cause clogs, jams, or excessive wear to plant machinery. In addition, the removal of various materials at the beginning of the treatment process saves valuable space within the treatment plant.

Preliminary treatment may include many different processes, each designed to remove a specific type of material that is a potential problem for the treatment process. Processes include wastewater collections—influent pumping (see Chapter 11), screening, shredding, grit removal, flow measurement, preaeration, chemical addition, and flow equalization. The major processes are shown in Figure 12.1. In this chapter, we describe and discuss each of these processes and their importance in the treatment process.

✓ Not all treatment plants will include all of the processes shown in Figure 12.1. Specific processes have been included to facilitate discussion of major potential problems with each process and its operation; this is information that may be important to the wastewater operator.

12.2 SCREENING

The purpose of screening is to remove large solids, such as rags, cans, rocks, branches, leaves, roots, etc., from the flow before the flow moves on to downstream processes.

✓ Typically, a treatment plant will remove anywhere from 0.5 to 12 ft^3 of screenings for each million gallons of influent received.

A *bar screen* traps debris as wastewater influent passes through. Typically, a bar screen consists of a series of parallel, evenly spaced bars (see Figure 12.2) or a perforated screen, placed in a channel. The wastestream passes through the screen, and the large solids (screenings) are trapped on the bars for removal.

✓ The screenings must be removed frequently enough to prevent accumulation, which will block the screen and cause the water level in front of the screen to build up.

The bar screen may be coarse (2 to 4-in. openings) or fine (0.75 to 2.0-in. openings). The bar screen may be manually cleaned (bars or screens are placed at an angle of 30° for easier solids removal—see

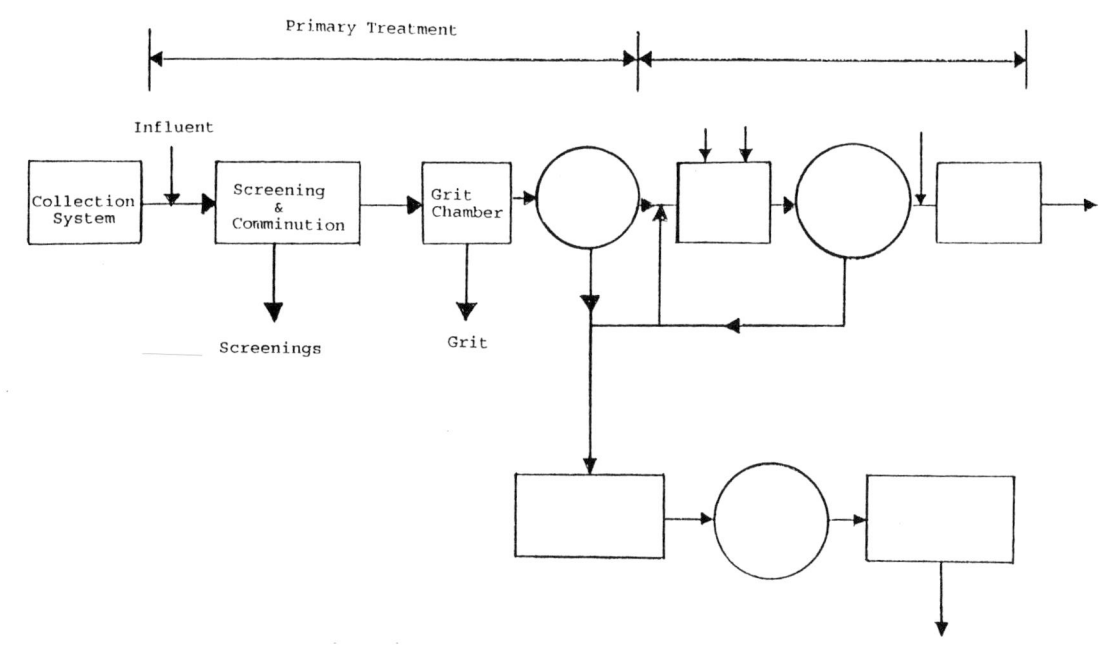

Figure 12.1 Preliminary treatment (unit processes).

Figure 12.2) or mechanically cleaned (bars are placed at a 45° to 60° angle to improve mechanical cleaner operation).

The screening method employed depends on the design of the plant, the amount of solids expected, and whether the screen is for constant or emergency use only.

12.2.1 MANUALLY CLEANED SCREENS

Manually cleaned screens are cleaned at least once per shift (or often enough to prevent buildup, which may cause reduced flow into the plant) using a long tooth rake. Solids are manually pulled to the drain platform and are allowed to drain before storage in a covered container.

The area around the screen should be cleaned frequently to prevent a buildup of grease or other

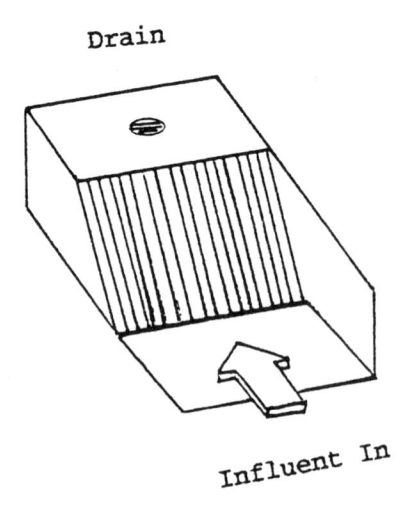

Figure 12.2 Manually cleaned bar screen.

materials that can cause odors, slippery conditions, and insect and rodent problems. Because screenings may contain organic matter as well as large amounts of grease, they should be stored in a covered container. Screenings can be disposed of by burial in approved landfills or by incineration. Some treatment facilities grind the screenings into small particles that are then returned to the wastewater flow for further processing and removal later in the process.

12.2.1.1 Operational Problems

Manually cleaned screens require a certain amount of operator attention to maintain optimum operation. Failure to clean the screen frequently can lead to septic wastes entering the primary; surge flows after cleaning; and/or low flows before cleaning. On occasion, when such operational problems occur, it becomes necessary to increase the frequency of the cleaning cycle. Another operational problem is excessive grit in the bar screen channel. This problem may be caused by improper design or construction or insufficient cleaning. The corrective action required is either to correct the design problem or increase cleaning frequency and flush the channel regularly. Another common problem with manually cleaned bar screens is their tendency to clog frequently. This may be caused by excessive debris in the wastewater or the screen is too fine for its current application. The operator should locate the source of the excessive debris and eliminate it. If the screen is the problem, a coarser screen may need to be installed. If the bar screen area is filled with obnoxious odors, flies, and other insects, it may be necessary to dispose of screening more frequently.

12.2.2 MECHANICALLY CLEANED SCREENS

Mechanically cleaned screens use a mechanized rake assembly to collect the solids and move them (carry them) out of the wastewater flow for discharge to a storage hopper. The screen may be continuously cleaned or cleaned on a time- or flow-controlled cycle. As with the manually cleaned screen, the area surrounding the mechanically operated screen must be cleaned frequently to prevent buildup of materials, which can cause unsafe conditions.

As with all mechanical equipment, operator vigilance is required to ensure proper operation and proper maintenance. Maintenance includes lubricating equipment and maintaining it in accordance with manufacturer's recommendations or the plant's operations and maintenance manual.

Screenings from mechanically operated bar screens are disposed of in the same manner as screenings from manually operated screens: landfill disposal, incineration, or ground for return to the wastewater flow.

12.2.2.1 Operational Problems

Many of the operational problems associated with mechanically cleaned bar screens are the same as those for manual screens: septic wastes entering the primary; surge flows after cleaning; excessive grit in the bar screen channel; and/or the screen clogs frequently. Basically, the same corrective actions employed for manually operated screens would be applied for these problems in mechanically operated screens. In addition to these problems, however, mechanically operated screens also have other problems, including the cleaner will not operate at all and the rake does not operate but the motor does. Obviously, these are mechanical problems that could be caused by a jammed cleaning mechanism, broken chain, broken cable, or a broken shear pin. Authorized and fully trained maintenance operators should be called in to handle these types of problems.

12.2.3 SAFETY

The screening area is the first location where the operator is exposed to the wastewater flow. Any toxic, flammable, or explosive gases present in the wastewater can be released at this point. Operators who frequent enclosed bar screen areas should be equipped with personal air monitors. Adequate ventilation must be provided. It is also important to remember that, due to the grease attached to the screenings, this area of the plant can be extremely slippery. Routine cleaning is required to minimize this problem.

✓ Never override safety devices on mechanical equipment. Overrides can result in dangerous conditions, injuries, and major mechanical failure.

12.3 SHREDDING

As an alternative to screening, shredding can be used to reduce solids to a size that can enter the plant without causing mechanical problems or clogging. Shredding processes include comminution (comminute means "cut up") and barminution devices.

12.3.1 COMMINUTION

The comminutor is the most common shredding device used in wastewater treatment. In this device, all the wastewater flow passes through the grinder assembly. The grinder consists of a screen or slotted basket, a rotating or oscillating cutter, and a stationary cutter. Solids pass through the screen and are chopped or shredded between the two cutters. The comminutor will not remove solids that are too large to fit through the slots, and it will not remove floating objects. These materials must be removed manually.

Maintenance requirements for comminutors include aligning, sharpening, and replacing cutters and corrective and preventive maintenance performed in accordance with the plant's operations and maintenance manual.

12.3.1.1 Operational Problems

Common operational problems associated with comminutors include output containing coarse solids. When this occurs, it is usually a sign that the cutters are dull or misaligned. If the system does not operate at all, the unit is either clogged or jammed, a shear pin or coupling is broken, or electrical power is shut off. If the unit stalls or jams frequently, this usually indicates cutter misalignment, excessive debris in influent, or dull cutters.

✓ Only qualified maintenance operators should perform maintenance of shredding equipment.

12.3.2 BARMINUTION

In barminution, the barminutor uses a bar screen to collect solids that are then shredded and passed through the bar screen for removal at a later process. In operation, each device cutter's alignment and sharpness are critical factors in effective operation. Cutters must be sharpened or replaced, and alignment must be checked in accordance with manufacturer's recommendations.

Solids that are not shredded must be removed daily, stored in closed containers, and disposed of by burial or incineration.

Barminutor operational problems are similar to those listed in Section 12.3.1.1. Preventive and corrective maintenance as well as lubrication must be performed by qualified personnel and in accordance with the plant's operations and maintenance manual. Because of higher maintenance requirements, the barminutor is less frequently used.

12.4 GRIT REMOVAL

The purpose of grit removal is to remove the heavy inorganic solids that could cause excessive mechanical wear. Grit is heavier than inorganic solids and includes sand, gravel, clay, egg shells, coffee grounds, metal filings, seeds, and other similar materials.

There are several processes or devices used for grit removal. All of the processes are based on the fact that grit is heavier than the organic solids, which should be kept in suspension for treatment in following processes. Grit removal may be accomplished in grit chambers or by the centrifugal separation of sludge. Processes use gravity/velocity, aeration, or centrifugal force to separate the solids from the wastewater.

12.4.1 GRAVITY/VELOCITY CONTROLLED GRIT REMOVAL

Gravity/velocity controlled grit removal is normally accomplished in a channel or tank where the speed or the velocity of the wastewater is controlled to about 1 ft per second (ideal)—so that grit will settle while organic matter remains suspended (see Figure 12.3). As long as the velocity is controlled in the range of 0.7 to 1.4 ft per second (fps), the grit removal will remain effective. Velocity is controlled by the amount of water flowing through the channel, the depth of the water in the channel, by the width of the channel, or by cumulative width of channels in service.

12.4.1.1 Process Control Calculations

Velocity of the flow in a channel can be determined either by the float and stopwatch method or by channel dimensions.

Example 12.1

Velocity by float and stopwatch.

$$\text{Velocity ft/sec} = \frac{\text{Distance Traveled, ft}}{\text{Time Required, s}} \qquad (12.1)$$

Problem:

A float takes 25 sec to travel 34 ft in a grit channel. What is the velocity of the flow in the channel?

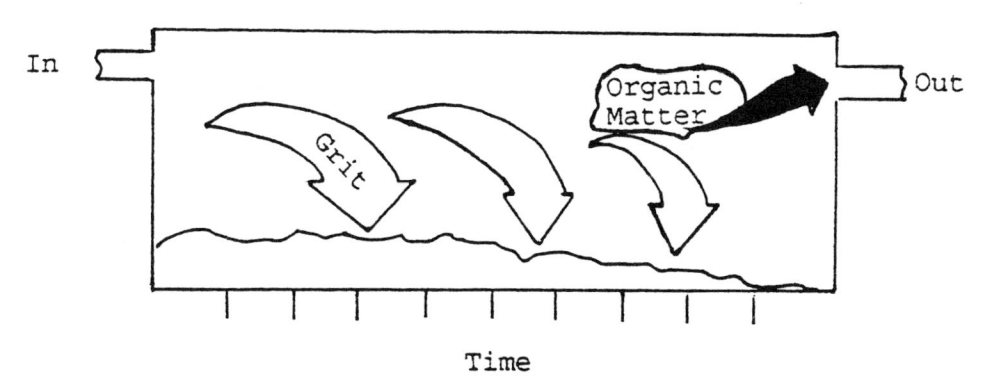

Figure 12.3 Gravity/velocity controlled grit removal.

Solution:

$$\text{Velocity, fps} = \frac{34 \text{ ft}}{25 \text{ s}} = 1.4 \text{ fps}$$

Example 12.2

Velocity by flow and channel dimensions.

✓ This calculation can be used for a single channel or tank or multiple channels or tanks with the same dimensions and equal flow. If the flow through each unit of the unit dimensions is unequal, the velocity for each channel or tank must be computed individually.

$$\text{Velocity, fps} = \frac{\text{Flow, MGD} \times 1.55 \text{ cfs/MGD}}{\text{\# Channels in Service} \times \text{Channel Width, ft} \times \text{Water Depth, ft}} \quad (12.2)$$

Problem:

The plant is currently using two grit channels. Each channel is 3 ft wide and has a water depth of 1.2 ft. What is the velocity when the influent flow rate is 3.0 MGD?

Solution:

$$\text{Velocity, fps} = \frac{3.0 \text{ MGD} \times 1.55 \text{ cfs/MGD}}{2 \text{ Channels} \times 3 \text{ ft} \times 1.2 \text{ ft}}$$

$$\text{Velocity, fps} = \frac{4.65 \text{ cfs}}{7.2 \text{ ft}^2} = 0.65 \text{ fps}$$

✓ *Note:* Since 0.65 is below the 0.7–1.4 level, the operator of this unit could consider taking one of the two channels out of service to increase the velocity to 1.30 fps, the range desired.

✓ The channel dimensions must always be in feet. Convert inches to feet by dividing by 12 in. per ft.

Example 12.3

Required settling time.

✓ This calculation can be used to determine the time required for a particle to travel from the surface of the liquid to the bottom at a given settling velocity. In order to compute the settling time, the settling velocity in fps must be provided or determined experimentally in a laboratory.

$$\text{Settling Time, sec} = \frac{\text{Liquid Depth in Feet}}{\text{Settling, Velocity, fps}} \quad (12.3)$$

Problem:

The plant's grit channel is designed to remove sand, which has a settling velocity of 0.085 fps.

The channel is currently operating at a depth of 2.2 ft. How many seconds will it take for a sand particle to reach the channel bottom?

Solution:

$$\text{Settling Time, sec} = \frac{2.2 \text{ ft}}{0.085 \text{ fps}} = 25.9 \text{ s}$$

Example 12.4

Required channel length.

✓ This calculation can be used to determine the length of a channel required to remove an object with a specified settling velocity.

$$\text{Required Channel Length} = \frac{\text{Channel Depth, ft} \times \text{Flow Velocity, fps}}{\text{Settling Velocity, fps}} \quad (12.4)$$

Problem:

The plant's grit channel is designed to remove sand, which has a settling velocity of 0.070 fps. The channel is currently operating at a depth of 3 ft. The calculated velocity of flow through the channel is 0.80 fps. The channel is 35 ft long. Is the channel long enough to remove the desired sand particle size?

Solution:

$$\text{Required Channel Length, ft} = \frac{3 \text{ ft} \times 0.80 \text{ fps}}{0.070 \text{ fps}} = 34.3 \text{ ft}$$

Yes, the channel is long enough to ensure all of the sand will be removed.

12.4.1.2 Cleaning

Gravity type systems may be manually or mechanically cleaned. Manual cleaning normally requires that the channel be taken out of service, drained, and manually cleaned. Mechanical cleaning systems are operated continuously or on a time cycle. Removal should be frequent enough to prevent grit carry-over into the rest of the plant.

✓ Before and during cleaning activities, always ventilate the area thoroughly.

12.4.2 AERATED SYSTEMS

Aerated grit removal systems use aeration to keep the lighter organic solids in suspension while allowing the heavier grit particles to settle out. Aerated grit removal may be manually or mechanically cleaned; however, the majority of the systems are mechanically cleaned.

In normal operation, the aeration rate is adjusted to produce the desired separation, which requires observation of mixing and aeration and sampling of fixed suspended solids. Actual grit removal is

12.4.3 CYCLONE DEGRITTER

The cyclone degritter uses a rapid spinning motion (centrifugal force) to separate the heavy inorganic solids or grit from the light organic solids. This unit process is normally used on primary sludge rather than the entire wastewater flow. The critical control factor for the process is the inlet pressure. If the pressure exceeds the recommendations of the manufacturer, the unit will flood, and grit will carry through with the flow. Grit is separated from the flow, washed, and discharged directly to a storage container. Grit removal performance is determined by calculating the percent removal for inorganic (fixed) suspended solids.

12.5 FLOW MEASUREMENT

Flow measurement is used throughout the treatment process to ensure the efficient operation of the treatment facility and to provide information (hydraulic and organic loading) needed to prepare required compliance reports for regulatory agencies.

There are many different methods available for measuring flows. The methodology used is generally based upon making a physical measurement that can then be related to the quantity of liquid moving past a given point in a specified length of time.

✓ For our purposes in this fundamental level presentation, we are concerned with the three methods that are currently most commonly used: fill and draw, weirs, and flumes.

12.5.1 FILL AND DRAW

In the fill and draw method of flow measurement, liquid flows into a container of known volume. The time required is measured. Dividing the liquid volume by time results in a flow rate.

The actual methodology used is explained as follows:

(1) Measure the dimensions of the container.
(2) Calculate the volume of the container.
(3) Determine the time required to collect the desired volume.
(4) Calculate the flow.

Let's take a look at exactly how the fill and draw method is used in practice.

Example 12.5

Problem:

The aerobic digester is 50 ft long and 25 ft wide. The waste sludge pump operates for 40 min and the level of the liquid in the digester increases 2.5 ft. What is the flow rate in gallons per minute?

Solution:

$$\text{Volume Pumped, ft}^3 = 50 \text{ ft} \times 25 \text{ ft} \times 2.5 \text{ ft}$$

$$= 3{,}125 \text{ ft}^3$$

$$\text{Volume Pumped, gal} = 3{,}125 \text{ ft}^3 \times 7.48 \text{ gal}/\text{ft}^3$$

$$= 23{,}375 \text{ gal}$$

$$\text{Flow Rate, gpm} = \frac{23{,}375 \text{ gal}}{40 \text{ min}}$$

$$= 548 \text{ gpm}$$

12.5.2 WEIRS

When a constriction or barrier is placed in an open channel, the amount of water that passes over or through the constriction is directly proportional to the head (height) of the water behind the constriction and the area of the opening in the constriction. The most widely used wastewater flow measurement devices for open channels are based upon this hydraulic principle. Weirs are one of these flow measurement devices. Because the area of the opening in the constriction remains constant, the only required measurement is the head behind the constriction. With this information, the flow rate can be calculated.

Let's take a closer look at weirs and how they actually operate.

Configured in many different forms, the basic design of the weir, in all cases, is a constriction or a dam placed across a channel. The most common type of weir used in wastewater treatment is the V-notch weir (see Figure 12.4). This weir consists of a solid vertical plate with a sharp crest and a V-notch (or triangular) cut in the top edge. The V-notch may have a 22.5°, 30°, 45°, 60°, or 90° angle.

Normal operation of the V-notch weir requires measurement of the head (distance from the surface of the water to the bottom of the V-notch at a point that is at least two times the maximum head of the weir behind the weir). The head over the weir must be measured accurately by a float, hook gauge, or level sensor. The head measurement is then converted to flow rate by reading a chart or by calculation. Due to the complexity of the calculation, the use of a chart or instrumentation that automatically converts the head reading is recommended. Weirs are accurate, simple to install, and relatively inexpensive. Their main disadvantages include large amounts of head loss and settling of solids upstream of the weir (Qasim, 1999). The following formula can be used to calculate the flow rate across a V-notch weir.

$$\text{Flow } (Q), \text{CFS} = K \times H^{2.5} \qquad (12.5)$$

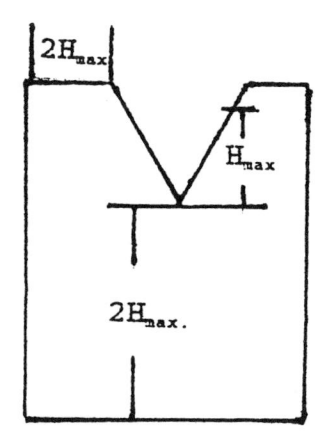

Figure 12.4 V-notch weir.

where

H = head in feet

K = constant related to weir angle

45° = 1.035

90° = 2.500

Weirs require periodic cleaning to ensure that the head area relationship is not changed by growth or buildup of debris in the notch. Weirs that use devices to sense, display, and record flow rates must be checked periodically (annually) by a qualified technician.

12.5.3 FLUMES

Flumes use the same basic hydraulic principle as the weir. The major difference is in the construction of the unit. The flume design includes a very narrow section known as the throat of the flume and a section that gradually changes from the width of the channel to the width of the throat (see Figure 12.5). This section is known as the converging section. The throat of the flume produces a head in the converging section that can be measured and converted to flow rate. Critical depth is measured at the flume.

There are several types of flumes. Probably the most common flume in wastewater treatment is the *Parshall Flume*. Normal operation of the Parshall Flume is the measurement of the head upstream of the throat at a point that is approximately two-thirds the length of the converging section. Head can be measured manually or by mechanical means (floats, sonic units, etc.). Flow rates can then be determined from a chart, by calculation, or automatically using mechanical or electronic methods. Due to the complexity of the calculation, flows are normally determined using a chart provided with the flume or by plant instrumentation.

✓ All plant instrumentation, including flow measurement devices, must be serviced periodically by a qualified instrument technician. The frequency of service is dependent on the quality of the instrument and the application for which it is used.

12.6 PREAERATION

In the preaeration process, wastewater is aerated to achieve and maintain an aerobic state (to freshen septic wastes), strip off hydrogen sulfide (to reduce odors and corrosion), and agitate solids to release trapped gases (improve solids separation and settling). All of this can be accomplished by aerating the wastewater for 10 to 30 minutes. To reduce BOD_5, preaeration must be conducted from 45 to 60 minutes.

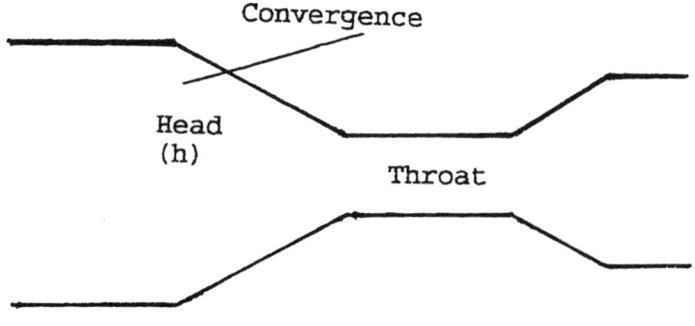

Figure 12.5 Parshall Flume.

12.7 CHEMICAL ADDITION

Chemical addition is made to the wastestream in order to reduce odors, neutralize acids or bases, reduce corrosion, reduce BOD$_5$, improve solids and grease removal, reduce loading on the plant, and to aid subsequent processes. Actual chemical use depends on the desired result. Chemicals must be added at a point where sufficient mixing will occur to obtain maximum benefit. Chemicals typically used in wastewater treatment include chlorine, peroxide, acids and bases, mineral salts (ferric chloride, alum, etc.), and bioadditives and enzymes.

12.8 FLOW EQUALIZATION

The purpose of flow equalization is to reduce or remove the wide swings in flow rates normally associated with wastewater treatment plant loadings. The process can be designed to prevent flows above maximum plant design hydraulic capacity; to reduce the magnitude of diurnal flow variations; and to eliminate flow variations. Flow equalization is accomplished using mixing or aeration equipment, pumps, and flow measurement. Normal operation depends on the purpose of the flow equalization system. Equalized flows allow the plant to perform at optimum levels by providing a stable hydraulic and organic loading. The downside to flow equalization is in the additional costs associated with construction and operation of the flow equalization facilities.

12.9 REFERENCE

Qasim, S. R., *Wastewater Treatment Plants: Planning, Design, and Operation,* 2nd ed. Lancaster, PA: Technomic Publishing Company, Inc., 1999.

12.10 CHAPTER REVIEW QUESTIONS

12-1 What is the purpose of preliminary treatment?

12-2 What is the purpose of the bar screen?

12-3 What two methods are available for cleaning a bar screen?

12-4 Name two ways to dispose of screenings.

12-5 What must be done to the cutters in a comminutor to ensure proper operation?

12-6 What controls the velocity in a gravity-type grit channel?

12-7 The plant has three channels in service. Each channel is 3 ft wide and has a water depth of 2 ft. What is the velocity in the channel when the flow rate is 8.0 MGD?

12-8 List three reasons why you might wish to include preaeration in the preliminary treatment portion of your plant.

12-9 Name two reasons why we would want to remove grit.

12-10 How slow should the flow of wastewater be in order to settle the grit?

12-11 Below what velocity will grit settle in the screening channel?

12-12 An empty screenings hopper 4 ft by 5 ft is filled to an even depth of 24 in. over the course of 84 hours. If the average plant flow rate was 4.5 MGD during this period, how many cubic feet of screenings were removed per million gallons of wastewater received?

12-13 The decomposition process that results in the production of methane gas is known as _____ decomposition.

12-14 A V-notch weir is normally used to measure_____.

12-15 If untreated organic wastes are discharged to a stream, the dissolved oxygen level of the stream will_____.

12-16 The main purpose of the grit chamber is to_____.

12-17 The main purpose of primary treatment is to_____.

CHAPTER 13

Sedimentation

13.1 INTRODUCTION

THE purpose of primary sedimentation (or clarification) is to remove settleable organic and flotable solids. Normally, each primary clarification unit can be expected to remove 90–95% settleable solids, 40–60% total suspended solids, and 25–35% BOD_5.

✓ *Note:* Performance expectations for settling devices used in other areas of plant operation are normally expressed as overall unit performance rather than settling unit performance.

Sedimentation may be used throughout the plant to remove settleable and flotable solids. It is used in primary treatment, secondary treatment, and advanced wastewater treatment processes. In this chapter, we focus on primary treatment or primary clarification, which uses large basins in which primary settling is achieved under relatively quiescent conditions (see Figure 13.1).

Within these basins, the primary settled solids are collected by mechanical scrapers into a hopper, from which they are pumped to a sludge-processing area. Oil, grease, and other floating materials (scum) are skimmed from the surface. The effluent is discharged over weirs into a collection trough.

13.2 PROCESS DESCRIPTION

In primary sedimentation, wastewater enters a settling tank or basin (see Figure 13.2). Velocity is reduced to approximately 1 ft per minute.

✓ Notice that the velocity is based on minutes instead of seconds as was the case in the grit channels. A grit channel velocity of 1 ft/s would be 60 ft/min.

Solids that are heavier than water settle to the bottom while solids that are lighter than water float to the top. Settled solids are removed as sludge, and floating solids are removed as scum. Wastewater leaves the sedimentation tank over an effluent weir and goes on to the next step in treatment. The efficiency of the process is controlled by detention time, temperature, tank design, and condition of the equipment.

13.3 TYPES OF SEDIMENTATION TANKS

Sedimentation equipment includes septic tanks, two-story tanks, and plain settling tanks or clarifiers. All three devices may be used for primary treatment, while plain settling tanks are normally used for secondary or advanced wastewater treatment processes.

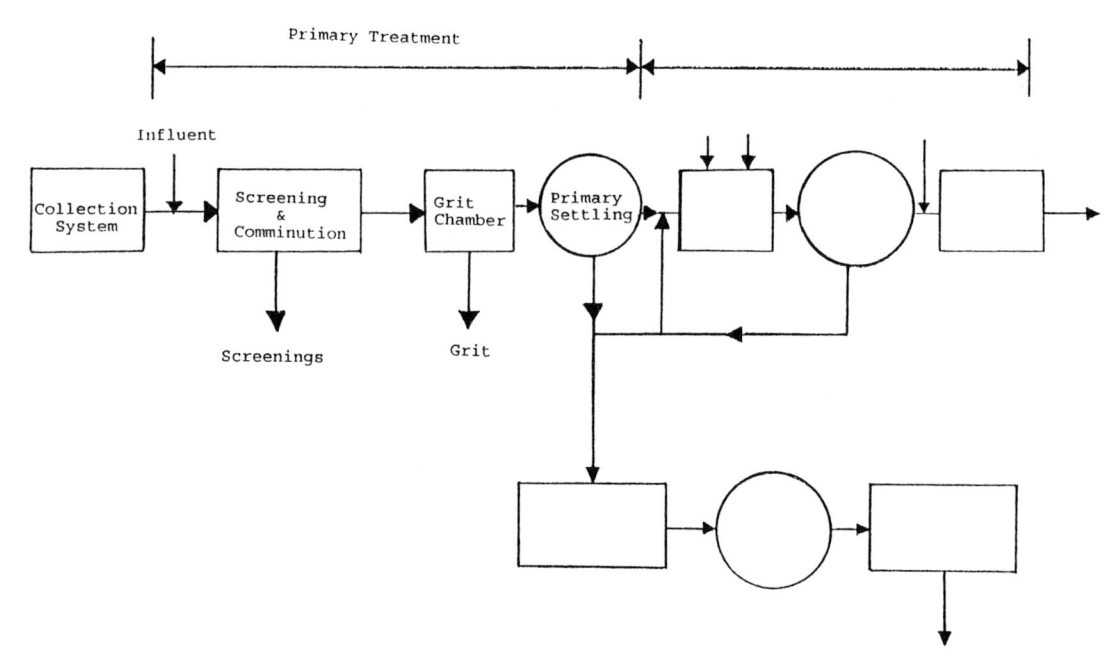

Figure 13.1 Sedimentation.

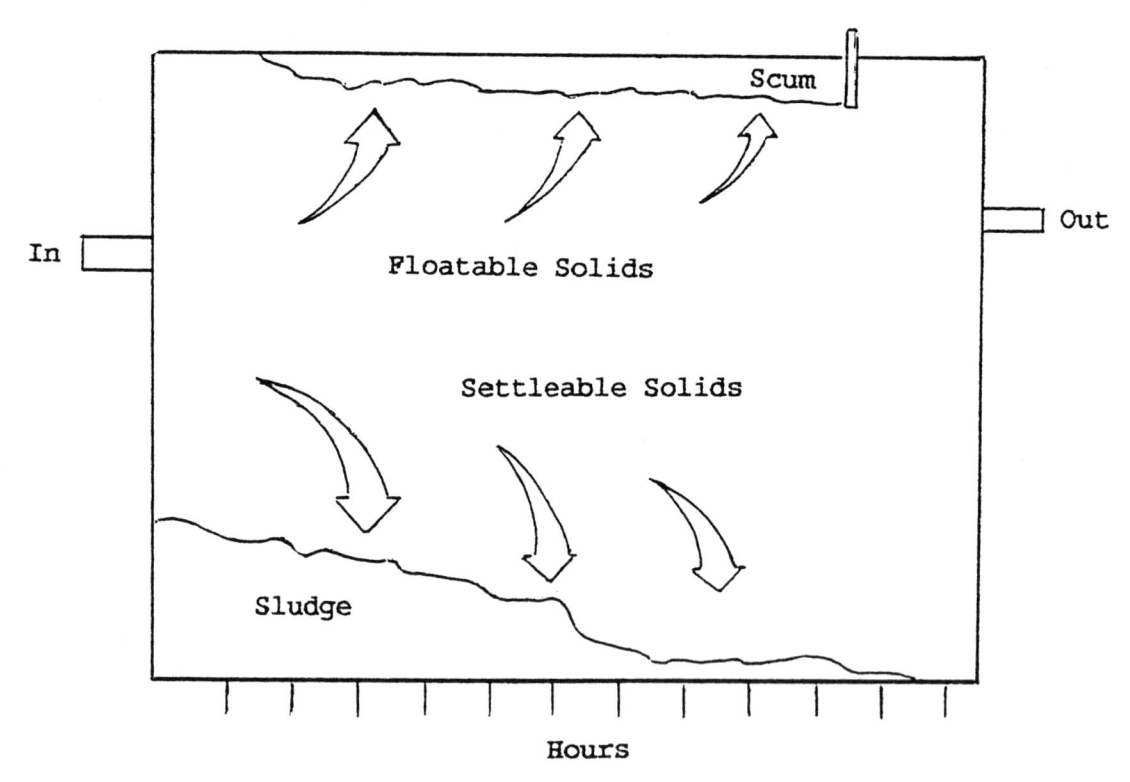

Figure 13.2 Sedimentation/flotation process.

13.3.1 SEPTIC TANKS

Septic tanks are prefabricated tanks (see Figure 13.3) that serve as a combined settling and skimming tank and as an unheated-unmixed anaerobic digester (Metcalf & Eddy, 1991). Septic tanks provide long settling times (6 to 8 hours or more), but do not separate decomposing solids from the wastewater flow. When the tank becomes full, solids will be discharged with the flow. The process is suitable for small facilities (i.e., schools, motels, homes, etc.) but, due to the long detention times and lack of control, it is not suitable for larger applications.

13.3.2 TWO-STORY (IMHOFF) TANK

The two-story or Imhoff tank is similar to a septic tank in the removal of settleable solids and the anaerobic digestion of solids. The difference is that the two-story tank consists of a settling compartment where sedimentation is accomplished, a lower compartment where settled solids and digestion take place, and gas vents (see Figure 13.4). Solids removed from the wastewater by settling pass from the settling compartment into the digestion compartment through a slot in the bottom of the settling compartment. The design of the slot prevents solids from returning to the settling compartment. Solids decompose anaerobically in the digestion section. Gases produced as a result of the solids decomposition are released through the gas vents running along each side of the settling compartment.

13.3.3 PLAIN SETTLING TANKS (CLARIFIERS)

The plain settling tank or clarifier optimizes the settling process. Sludge is removed from the tank for processing in other downstream treatment units. Flow enters the tank, is slowed and distributed evenly across the width and depth of the unit, passes through the unit, and leaves over the effluent weir (see Figure 13.5). Detention time within the primary settling tank is from 1–3 hours (2-hour average).

Sludge removal is accomplished frequently on either a continuous or intermittent basis. Continuous removal requires additional sludge treatment processes to remove the excess water resulting from removal of sludge, which contains less than 2–3% solids. Intermittent sludge removal requires the sludge to be pumped from the tank on a schedule frequent enough to prevent large clumps of solids rising to the surface but infrequent enough to obtain 4–8% solids in the sludge withdrawn.

Scum must be removed from the surface of the settling tank frequently. This is normally a mechanical process but may require manual start-up. The system should be operated frequently enough to prevent excessive buildup and scum carryover but not so frequently as to cause hydraulic overloading of the scum removal system.

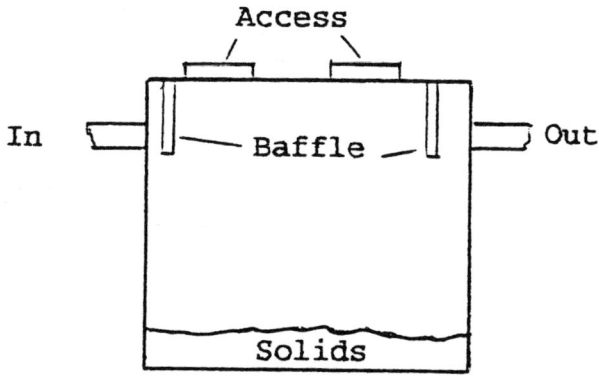

Figure 13.3 Septic tank.

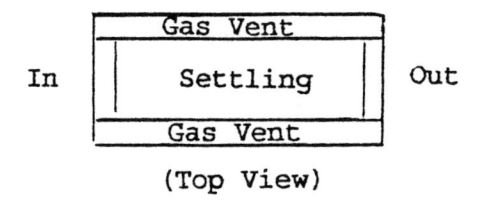

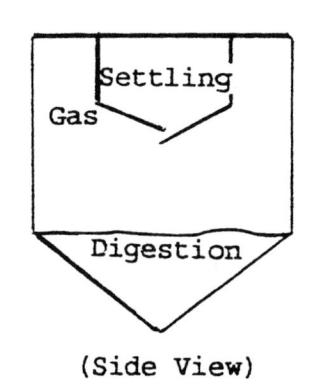

Figure 13.4 Two-story (Imhoff) tank.

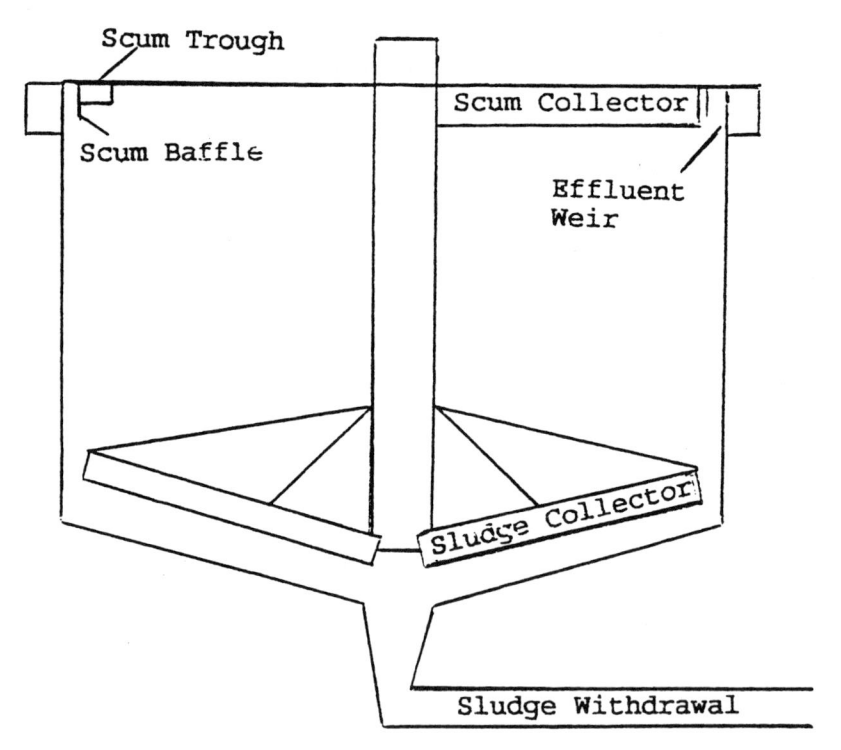

Figure 13.5 Plain settling tank.

Settling tanks require housekeeping and maintenance. Baffles (prevent flotable solids, scum, from leaving the tank), scum troughs, scum collectors, effluent troughs, and effluent weirs require frequent cleaning to prevent heavy biological growths and solids accumulations. Mechanical equipment must be lubricated and maintained as specified in the manufacturer's recommendations or in accordance with procedures listed in the plant's operations and maintenance manual.

Process control sampling and testing is used to evaluate the performance of the settling process. Settleable solids, dissolved oxygen, pH, temperature, total suspended solids, and BOD_5, as well as sludge solids and volatile matter testing, are routinely accomplished.

13.4 PROCESS CONTROL CALCULATIONS

As with many other wastewater treatment plant unit processes, process control calculations aid in determining the performance of the sedimentation process. Process control calculations are used in the sedimentation process to determine

- percent removal (see Chapter 5, Section 5.3)
- hydraulic detention time (see Chapter 6)
- surface loading rate (surface settling rate)
- weir overflow rate (weir loading rate)
- sludge pumping
- percent total solids (% TS)

In the following sections, we take a closer look at a few of these process control calculations.

13.4.1 SURFACE LOADING RATE (SURFACE SETTLING RATE)

Surface loading rate is the number of gallons of wastewater passing over 1 square foot of tank per day. This can be used to compare actual conditions with design. Plant designs generally use a surface loading rate of 300 to 1,200 gal/day/ft^2.

$$\text{Surface Settling Rate, gpd/ft}^2 = \frac{\text{Flow, gal/day}}{\text{Settling Tank Area, ft}^2} \qquad (13.1)$$

Example 13.1

Problem:

The settling tank is 120 ft in diameter, and the flow to the unit is 4.5 MGD. What is the surface loading rate in gal/day/ft^2?

Solution:

$$\text{Surface Loading Rate} = \frac{4.5 \text{ MGD} \times 1{,}000{,}000 \text{ gal/MGD}}{0.785 \times 120 \text{ ft} \times 120 \text{ ft}} = 398 \text{ gpd/ft}^2$$

13.4.2 WEIR OVERFLOW RATE (WEIR LOADING RATE)

Weir overflow rate (weir loading rate) is the amount of water leaving the settling tank per linear foot of weir. The result of this calculation can be compared with design. Normally, weir overflow rates of 10,000 to 20,000 gal/day/ft are used in the design of a settling tank.

$$\text{Weir Overflow Rate, gpd/ft} = \frac{\text{Flow, gal/day}}{\text{Weir Length, ft}} \quad (13.2)$$

Example 13.2

Problem:

The circular settling tank is 90 ft in diameter and has a weir along its circumference. The effluent flow rate is 2.55 MGD. What is the weir overflow rate in gallons per day per foot?

Solution:

$$\text{Weir Overflow, gpd/ft} = \frac{2.55 \text{ MGD} \times 1,000,000 \text{ gal/MG}}{3.14 \times 90 \text{ ft}}$$

$$= 9,023 \text{ gal/day/ft}$$

✓ *Note:* Notice that 9,023 gal/day/ft is below the recommended minimum of 10,000.

13.4.3 SLUDGE PUMPING

Determination of sludge pumping (the quantity of solids and volatile solids removed from the sedimentation tank) provides accurate information needed for process control of the sedimentation process.

(1) Solids Pumped, lb/day = (13.3)

  Pump Rate, gpm × Pump Time, min/day × 8.34 lb/gal × % Solids

(2) Volatile Solids, lb/day = (13.4)

  Pump Rate × Pump Time × 8.34 × % Solids × % Volatile Matter

Example 13.3

Problem:

The sludge pump operates 20 minutes per hour. The pump delivers 20 gal/min of sludge. Laboratory tests indicate that the sludge is 5.2% solids and 66% volatile matter. How many pounds of volatile matter are transferred from the settling tank to the digester? Assume a 24-hour period.

Solution:

$$\text{Pump Time} = 20 \text{ min/h}$$
$$\text{Pump Rate} = 20 \text{ gpm}$$
$$\% \text{ Solids} = 5.2\%$$
$$\% \text{ VM} = 66\%$$
$$\text{Volatile Solids, lb/day} = 20 \text{ gpm} \times (20 \text{ min/h} \times 24 \text{ h/day}) \times 8.34 \text{ lb/gal} \times 0.052 \times 0.66$$
$$= 2,748 \text{ lb/day}$$

13.4.4 PERCENT TOTAL SOLIDS (% TS)

Example 13.4

Problem:

A settling tank sludge sample is tested for solids. The sample and dish weighed 74.69 g. The dish alone weighed 21.2 g. After drying, the dish with dry solids weighed 22.3 g. What is the percent total solids (% TS) of the sample?

Solution:

Sample + Dish	74.69 g	Dish + Dry Solids	22.3 g
Dish Alone	−21.2 g	Dish Alone	−21.2 g
Sample Weight	53.49 g	Dry Solids Weight	1.1 g

$$\frac{1.1 \text{ g}}{53.49 \text{ g}} \times 100\% = 2\%$$

13.5 EFFLUENT FROM SETTLING TANKS

Upon completion of screening, degritting, and settling in sedimentation basins, large debris, grit, and many settleable materials have been removed from the wastestream. What is left is referred to as primary effluent. Usually cloudy and frequently grey in color, primary effluent still contains large amounts of dissolved food and other chemicals (nutrients). These nutrients are treated in the next step in the treatment process (secondary treatment), which is discussed in detail in the next chapter.

13.6 REFERENCE

Metcalf & Eddy, *Wastewater Engineering: Treatment, Disposal, Reuse*, 3rd ed., New York: McGraw-Hill, Inc., 1991.

13.7 CHAPTER REVIEW QUESTIONS

13-1 What is the purpose of sedimentation?

13-2 The sludge pump operates 20 minutes every 3 hours. The pump delivers 75 gpm. If the sludge is 5.5% solids and has a volatile matter content of 66%, how many pounds of volatile solids are removed from the settling tank each day?

13-3 The circular settling tank is 80 ft in diameter and has a depth of 12 ft. The flow rate is 2.6 MGD. What is the detention time in hours, surface loading rate in gal/day/ft^2, and weir overflow rate in gal/day/ft?

13-4 What is the recommended procedure to follow when removing sludge intermittently from a primary settling tank?

13-5 Why is there normally a baffle at the effluent end of the primary settling tank?

13-6 How much of the settleable solids are removed by primary settling?

13-7 What is an average detention time in a primary clarifier?

13-8 A settling tank is 80 ft long, 20 ft wide, and 10 ft deep and receives a flow rate of 1.5 MGD. What is the surface overflow rate in gpd/ft^2?

13-9 A settling tank with a total weir length of 80 ft receives a flow rate of 1.25 MGD. What is the weir overflow rate in gpd/ft?

13-10 A wastewater treatment plant has six primary tanks. Each tank is 80 ft long, 20 ft wide with a side water depth of 12 ft, and a total weir length of 86 ft. The flow rate to the plant is 5 MGD. There are three tanks currently in service. Calculate the detention time in minutes, the surface overflow rate in gpd/ft^2, and the weir overflow rate in gpd/ft.

13-11 A primary settling tank is 80 ft in diameter and 10 ft deep. What is the detention time when the flow rate is 3.25 MGD?

CHAPTER 14

Secondary Treatment: Ponds, Trickling Filters, and RBCs

14.1 INTRODUCTION

THE main purpose of secondary treatment (sometimes referred to as biological treatment) is to provide biochemical oxygen demand (BOD) removal beyond what is achievable by primary treatment. There are three commonly used approaches, all of which take advantage of the ability of microorganisms to convert organic wastes (via biological treatment) into stabilized, low-energy compounds. Two of these approaches, the trickling filter [and/or its variation, the rotating biological contactor (RBC)] and the activated sludge process, sequentially follow normal primary treatment. The third, ponds (oxidation ponds or lagoons), however, can provide equivalent results without preliminary treatment.

In this chapter, a brief overview of the secondary treatment process followed by a detailed discussion of wastewater treatment ponds (used primarily in smaller treatment plants), trickling filters, and RBCs is presented. The focus then shifts to the activated sludge process (see Chapter 15)—the secondary treatment process, which is used primarily in large installations and is the main focus of the handbook.

14.2 SECONDARY TREATMENT

Secondary treatment refers to those treatment processes that use biological processes to convert dissolved, suspended, and colloidal organic wastes to more stable solids, which can either be removed by settling or discharged to the environment without causing harm.

Exactly what is secondary treatment? As defined by the Clean Water Act (CWA), secondary treatment produces an effluent with no more than 30 mg/L BOD_5 and 30 mg/L total suspended solids.

✓ The CWA also states that ponds and trickling filters will be included in the definition of secondary treatment even if they do not meet the effluent quality requirements continuously.

Most secondary treatment processes decompose solids aerobically producing carbon dioxide, stable solids, and more organisms. Because solids are produced, all of the biological processes must include some form of solids removal (settling tank, filter, etc.).

Secondary treatment processes can be separated into two large categories: fixed film systems and suspended growth systems.

Fixed film systems are processes that use a biological growth (biomass or slime) that is attached to some form of media. Wastewater passes over or around the media and the slime. When the wastewater and slime are in contact, the organisms remove and oxidize the organic solids. The media may be stone, redwood, synthetic materials, or any other substance that is durable (capable of withstanding weather conditions for many years), provides a large area for slime growth while providing open space for ventilation, and is not toxic to the organisms in the biomass. Fixed film devices include trickling filters and RBCs.

Suspended growth systems are processes that use a biological growth mixed with the wastewater. Typical suspended growth systems are the various modifications of the activated sludge process (see Chapter 15).

14.3 TREATMENT PONDS

Wastewater treatment can be accomplished using ponds. Ponds are relatively easy to build and to manage. They accommodate large fluctuations in flow, and can also provide treatment that approaches conventional systems (producing a highly purified effluent) at much lower cost. It is the cost (the economics) that drives many managers to decide on the pond option. The actual degree of treatment provided depends on the type and number of ponds used. Ponds can be used as the sole type of treatment, or they can be used in conjunction with other forms of wastewater treatment—that is, other treatment processes followed by a pond or a pond followed by other treatment processes.

14.3.1 TYPES OF PONDS

Ponds can be classified (named) based upon their location in the system, by the type of wastes they receive, and by the main biological process occurring in the pond. First, we look at the types of ponds according to their location and the type of wastes they receive: raw sewage stabilization ponds (see Figure 14.1), oxidation ponds, and polishing ponds. Then, in the following section, we look at ponds classified by the type of processes occurring within the pond: aerobic ponds, anaerobic ponds, facultative ponds, and aerated ponds.

14.3.1.1 Ponds Based on Location and Types of Wastes They Receive

14.3.1.1.1 Raw Sewage Stabilization Pond

The raw sewage stabilization pond is the most common type of pond (see Figure 14.1). With the exception of screening and shredding, this type of pond receives no prior treatment. Generally, raw sewage stabilization ponds are designed to provide a minimum of 45 days of detention time and to receive no more than 30 lb of BOD_5 per day per acre. The quality of the discharge is dependent on the time of the year. Summer months produce high BOD_5 removal but excellent suspended solids removals.

The pond consists of an influent structure, pond berm or walls, and an effluent structure designed to permit selection of the best quality effluent. Normal operating depth of the pond is 3 to 5 ft.

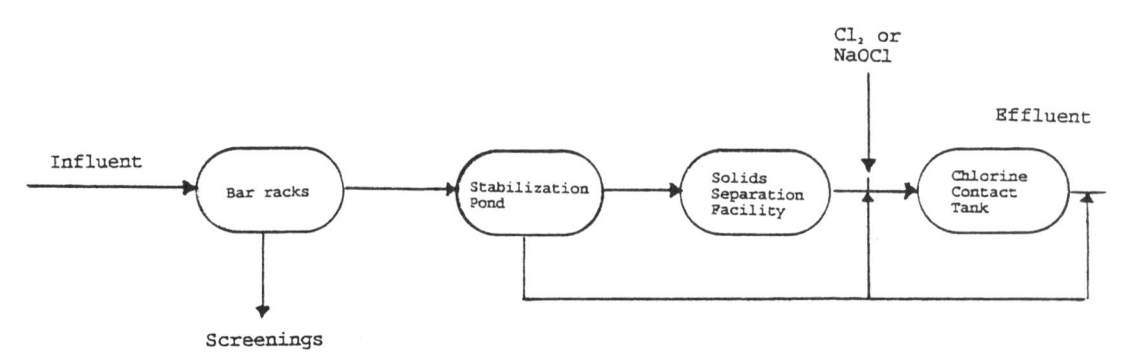

Figure 14.1 Typical (simplified) flow diagram for a stabilization pond.

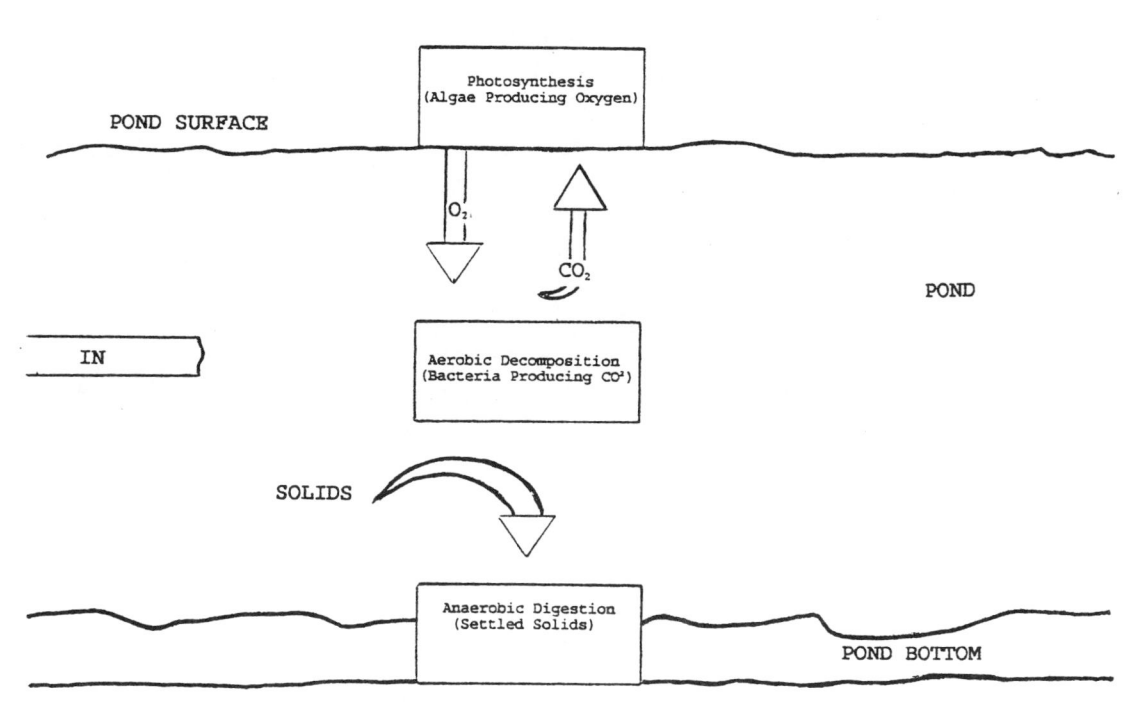

Figure 14.2 Stabilization pond processes.

The process occurring in the pond involves bacteria decomposing the organics in the wastewater (aerobically and anaerobically) and algae using the products of the bacterial action to produce oxygen (photosynthesis)—see Figure 14.2. Because this type of pond is the most commonly used in wastewater treatment, the process that occurs within the pond is described in greater detail in the following.

When wastewater enters the stabilization pond, several processes begin to occur. These include settling, aerobic decomposition, anaerobic decomposition, and photosynthesis. Solids in the wastewater will settle to the bottom of the pond. In addition to the solids in the wastewater entering the pond, solids that are produced by the biological activity will also settle to the bottom. Eventually, this will reduce the detention time and the performance of the pond. When this occurs (normally, 20–30 years), the pond will have to be replaced or cleaned.

Bacteria and other microorganisms use the organic matter as a food source. They use oxygen (aerobic decomposition), organic matter, and nutrients to produce carbon dioxide, water, stable solids, which may settle out, and more organisms. The carbon dioxide is an essential component of the photosynthesis process occurring near the surface of the pond.

Organisms also use the solids that settled out as food material; however, the oxygen levels at the bottom of the pond are extremely low, so the process used is anaerobic decomposition. The organisms use the organic matter to produce gases (hydrogen sulfide, methane, etc.), which are dissolved in the water, produce stable solids, and more organisms.

Near the surface of the pond, a population of green algae will develop, which can use the carbon dioxide produced by the bacterial population, nutrients, and sunlight to produce more algae and oxygen, which is dissolved into the water. The dissolved oxygen is then used by organisms in the aerobic decomposition process.

When compared with other wastewater treatment systems involving biological treatment, a stabilization pond treatment system is the simplest to operate and maintain. Operation and maintenance activities include collecting and testing samples for dissolved oxygen (DO) and pH, removing weeds and other debris (scum) from the pond, mowing the berms, repairing erosion, and removing burrowing animals.

✓ *Note:* Dissolved oxygen and pH levels in the pond will vary throughout the day. Normal operation will result in very high DO and pH levels due to the natural processes occurring.

✓ *Note:* When operating properly, the stabilization pond will exhibit a wide variation in both dissolved oxygen and pH. This is due to the photosynthesis occurring in the system.

14.3.1.1.2 Oxidation Pond

An oxidation pond, which is normally designed using the same criteria as the stabilization pond, receives flows that have passed through a stabilization pond or primary settling tank. This type of pond provides biological treatment, additional settling, and some reduction in the number of fecal coliform present.

14.3.1.1.3 Polishing Pond

A polishing pond, which uses the same equipment as a stabilization pond, receives flow from an oxidation pond or from other secondary treatment systems. Polishing ponds remove additional BOD_5, solids and fecal coliform, and some nutrients. They are designed to provide one to three days of detention time and normally operate at a depth of 5 to 10 feet. Excessive detention time or too shallow a depth will result in algae growth that increases influent suspended solids concentrations.

14.3.1.2 Ponds Based on the Type of Processes Occurring Within

Ponds may also be classified by the type of processes occurring within the pond. These include the aerobic, anaerobic, facultative, and aerated processes.

14.3.1.2.1 Aerobic Ponds

In aerobic ponds, which are not widely used, oxygen is present throughout the pond. All biological activity is aerobic decomposition.

14.3.1.2.2 Anaerobic Ponds

Anaerobic ponds are normally used to treat high-strength industrial wastes. No oxygen is present in the pond, and all biological activity is anaerobic decomposition.

14.3.1.2.3 Facultative Pond

The facultative pond is the most common type of pond (based on processes occurring). Oxygen is present in the upper portions of the pond, and aerobic processes are occurring. No oxygen is present in the lower levels of the pond where processes occurring are anoxic and anaerobic.

14.3.1.2.4 Aerated Pond

In the aerated pond, oxygen is provided through the use of mechanical or diffused air systems. When aeration is used, the depth of the pond and/or the acceptable loading levels may increase. Mechanical or diffused aeration is often used to supplement natural oxygen production or to replace it.

14.3.2 PROCESS CONTROL CALCULATIONS (STABILIZATION POND)

✓ *Note:* Process control calculations are an important part of wastewater treatment operations, including pond operations. More significantly, process control calculations are an important part of state wastewater licensing examinations—you simply cannot master the licensing examinations without being able to perform the required calculations. Thus, as with previous chapters (and with chapters to follow), whenever possible, example process control problems are provided to enhance your knowledge and skill.

14.3.2.1 Determining Pond Area in Acres

$$\text{Area, acres} = \frac{\text{Area, ft}^2}{43{,}560 \text{ ft}^2/\text{acre}} \tag{14.1}$$

14.3.2.2 Determining Pond Volume in Acre-Feet (ac-ft)

$$\text{Volume, ac-ft} = \frac{\text{Volume, ft}^3}{\text{Volume, ac-ft}} \tag{14.2}$$

14.3.2.3 Determining Flow Rate in Ac-Ft/Day

$$\text{Flow, acre-ft/day} = \text{Flow, MGD} \times 3.069 \text{ ac-ft/MG} \tag{14.3}$$

✓ *Note:* Acre-feet is a unit that can cause confusion, especially for those not familiar with pond or lagoon operations. One ac-ft is the volume of a box with a 1-acre top and 1 ft of depth—but, the top doesn't have to be an even number of acres in size to use acre-feet.

14.3.2.4 Determining Flow Rate in Acre-Inches/Day

$$\text{Flow, acre-inches/day} = \text{Flow, MGD} \times 36.8 \text{ ac-inches/MG} \tag{14.4}$$

14.3.2.5 Hydraulic Detention Time, Days

$$\text{Hydraulic Detention Time, Days} = \frac{\text{Pond Volume, ac-ft}}{\text{Influent Flow, ac-ft/day}} \tag{14.5}$$

✓ *Note:* Normally, hydraulic detention time ranges from 30 to 120 days for stabilization ponds.

Let's take a look at an example of how to determine detention time in days for a stabilization pond.

Example 14.1

Problem:

A stabilization pond has a volume of 53.5 ac-ft. What is the detention time in days when the flow is 0.30 MGD?

Solution:

$$\text{Flow, ac-ft/day} = 0.30 \text{ MGD} \times 3.069$$

$$= 0.92 \text{ ac-ft/day}$$

$$\text{Detention Time in Days} = \frac{53.5 \text{ acre}}{0.92 \text{ ac-ft/day}} = 58.2 \text{ days}$$

14.3.2.6 Hydraulic Loading, Inches/Day (Overflow Rate)

$$\text{Hydraulic Loading, inches/day} = \frac{\text{Influent Flow, acre-inches/day}}{\text{Pond Area, acres}} \quad (14.6)$$

14.3.2.7 Population Loading

$$\text{Population Loading, people/acre/day} = \frac{\text{Pop. Served by System, People}}{\text{Pond Area, acres}} \quad (14.7)$$

✓ *Note:* Population loading normally ranges from 50 to 500 people per acre.

14.3.2.8 Organic Loading

Organic loading can be expressed as pounds of BOD_5 per acre per day (most common), pounds BOD_5 per acre-foot per day, or people per acre per day.

$$\text{Organic L, lb } BOD_5/\text{acre/day} = \frac{BOD_5, \text{mg/L Infl. flow, MGD} \times 8.34}{\text{Pond Area, acres}} \quad (14.8)$$

✓ *Note:* Normal range is 10 to 50 lb BOD_5 per day per acre.

Example 14.2

Problem:

A wastewater treatment pond has an average width of 380 ft and an average length of 725 ft. The influent flow rate to the pond is 0.12 MGD with a BOD concentration of 160 mg/L. What is the organic loading rate to the pond in pounds per day per acre (lb/d/ac)?

Solution:

$$725 \text{ ft} \times 380 \text{ ft} \times \frac{1 \text{ acre}}{43,560 \text{ ft}^2} = 6.32 \text{ acre}$$

$$0.12 \text{ MGD} \times 160 \text{ mg/L} \times 8.34 \text{ lb/gal} = 160.1 \text{ lb/day}$$

$$\frac{160.1 \text{ lb/d}}{6.32 \text{ acre}} = 25.3 \text{ lb/d/ac}$$

14.4 TRICKLING FILTERS

In most wastewater treatment systems, the trickling filter follows primary treatment and includes a secondary settling tank or clarifier as shown in Figure 14.3. Trickling filters are widely used for the treatment of domestic and industrial wastes. The process is a fixed film biological treatment method designed to remove BOD_5 and suspended solids.

A trickling filter consists of a rotating distribution arm that sprays and evenly distributes liquid wastewater over a circular bed of fist-sized rocks, other coarse materials, or synthetic media (see Figure 14.4). The spaces between the media allow air to circulate easily so that aerobic conditions can be maintained. The spaces also allow wastewater to trickle down through, around, and over the media. The media material is covered by a layer of biological slime that absorbs and consumes the wastes trickling through the bed. The organisms aerobically decompose the solids, producing more organisms and stable wastes that either become part of the slime or are discharged back into the wastewater flowing over the media. This slime consists mainly of bacteria, but it may also include algae, protozoa, worms, snails, fungi, and insect larvae. The accumulating slime occasionally sloughs off (sloughings) individual media materials (see Figure 14.5) and is collected at the bottom of the filter, along with the treated wastewater, and is passed on to the secondary settling tank where it is removed.

The overall performance of the trickling filter is dependent on hydraulic and organic loading, temperature, and recirculation.

Let's take a closer look at the equipment that makes up a trickling filter.

14.4.1 TRICKLING FILTER EQUIPMENT

The distribution system is designed to spread wastewater evenly over the surface of the entire media. The most common system is the rotary distributor, which moves above the surface of the media and sprays the wastewater on the surface. The rotary system is driven by the force of the water leaving the orifices. The distributor arms usually have small plates below each orifice to spread the wastewater into a fan-shaped distribution system. The second type of distributor is the fixed nozzle system. In this system, the nozzles are fixed in place above the media and are designed to spray the wastewater over a fixed portion of the media. This system is used frequently with deep bed synthetic media filters.

✓ *Note:* Trickling filters that use ordinary rock are normally only about 10 ft in depth because of structural problems caused by the weight of rocks—which also requires the construction of beds that are quite wide, in many applications, up to 60 ft in diameter. When synthetic media are used, the bed can be much deeper.

No matter which type of media is selected, the primary consideration is that it must be capable of providing the desired film location for the development of the biomass. Depending on the type of media used and the filter classification, the media may be 3 to 20 or more feet in depth.

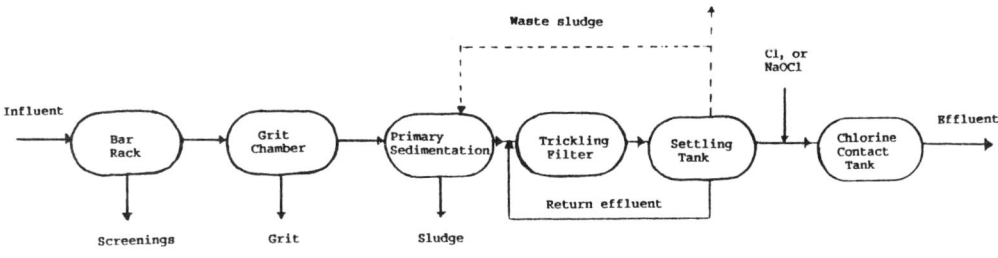

Figure 14.3 Simplified flow diagram of trickling filter used for wastewater treatment.

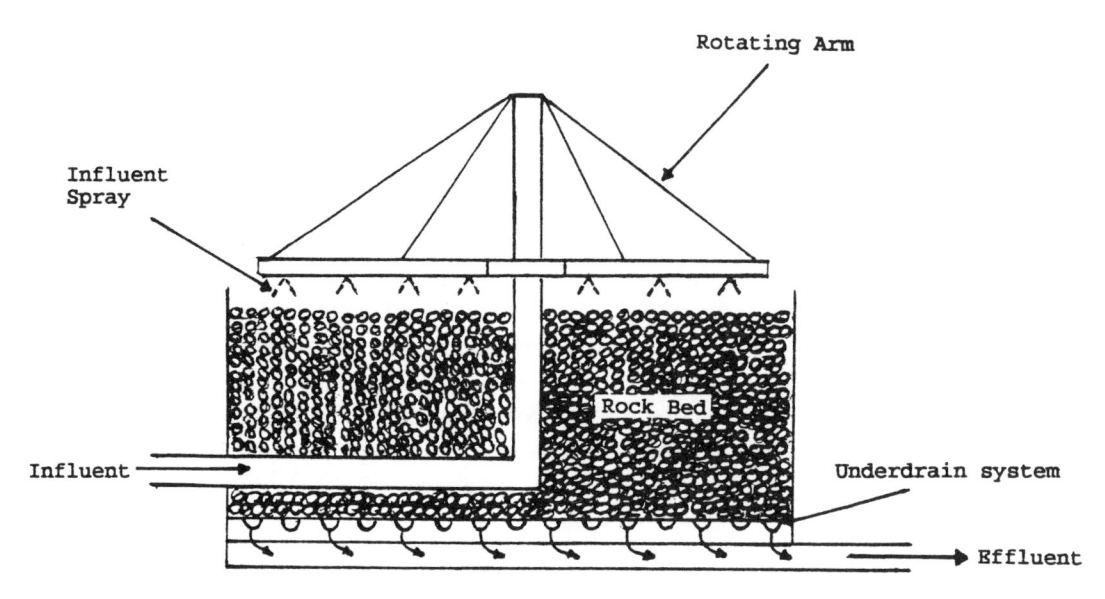

Figure 14.4 Schematic of the cross section of a trickling filter.

The underdrains are designed to support the media, collect the wastewater and sloughings, carry them out of the filter, and provide ventilation to the filter.

✓ In order to ensure sufficient air flow to the filter, the underdrains should never be allowed to flow more than 50% full of wastewater.

The effluent channel is designed to carry the flow from the trickling filter to the secondary settling tank.

The secondary settling tank provides 2 to 4 hours of detention time to separate the sloughing

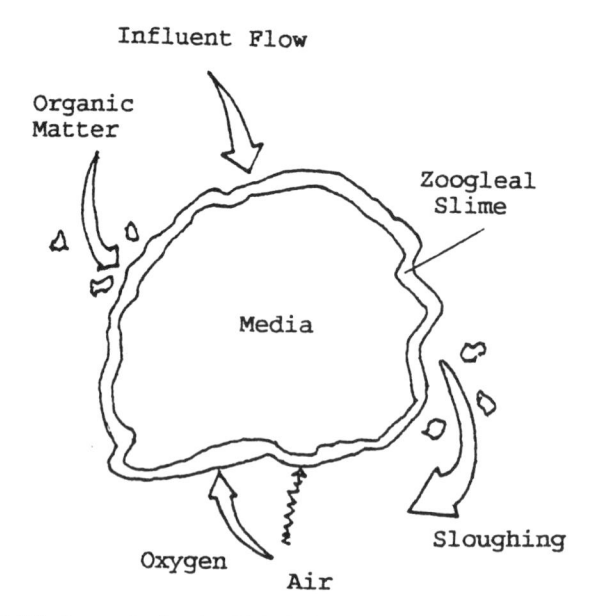

Figure 14.5 Filtering media showing biological activities that take place on surface area.

materials from the treated wastewater. Design, construction, and operation are similar to that of the primary settling tank. Longer detention times are provided because the sloughing materials are lighter and settle more slowly.

Recirculation pumps and piping are designed to recirculate (and thus improve the performance of the trickling filter or settling tank) a portion of the effluent back to be mixed with the filter influent. When recirculation is used, obviously, pumps and metering devices must be provided.

14.4.2 FILTER CLASSIFICATIONS

Trickling filters are classified by hydraulic and organic loading. Moreover, the expected performance and the construction of the trickling filter are determined by the filter classification. Filter classifications include standard rate, intermediate rate, high rate, super high rate (plastic media), and roughing rate types. Standard rate, high rate, and roughing rate are the filter types most commonly used.

The standard rate filter has a hydraulic loading (gpd/ft^3) of 25 to 90; a seasonal sloughing frequency; does not employ recirculation; and typically has an 80–85% BOD_5 removal rate and 80–85% TSS removal rate.

The high rate filter has a hydraulic loading (gpd/ft^3) of 230 to 900; a continuous sloughing frequency; always employs recirculation; and typically has an 65–80% BOD_5 removal rate and 65–80% TSS removal rate.

The roughing filter has a hydraulic loading (gpd/ft^3) of >900; a continuous sloughing frequency; does not normally include recirculation; and typically has a 40–65% removal rate and 40–65% TSS removal rate.

14.4.3 STANDARD OPERATING PROCEDURES

Standard operating procedures for trickling filters include sampling and testing, observation, recirculation, maintenance, and expectations of performance.

Collection of influent and process effluent samples to determine performance and monitor process condition of trickling filters is required. Dissolved oxygen, pH, and settleable solids testing should be collected daily. BOD_5 and suspended solids testing should be done as often as practical to determine the percent removal.

The operation and condition of the filter should be observed daily. Items to observe include the distributor movement, uniformity of distribution, evidence of operation or mechanical problems, and the presence of objectionable odors. In addition to the items above, the normal observation for a settling tank should also be performed.

Recirculation is used to reduce the organic loading, improve sloughing, reduce odors, and reduce or eliminate filter fly or ponding problems. The amount of recirculation is dependent on the design of the treatment plant and the operational requirements of the process. Recirculation flow may be expressed as a specific flow rate (i.e., 2.0 MGD). In most cases, it is expressed as a ratio (3:1, 0.5:1.0, etc.). The recirculation is always listed as the first number, and the influent flow is listed as the second number.

✓ Because the second number in the ratio is always 1.0, the ratio is sometimes written as a single number (dropping the :1.0)

Flows can be recirculated from various points following the filter to various points before the filter. The most common form of recirculation removes flow from the filter effluent or settling tank and returns it to the influent of the trickling filter as shown in Figure 14.6.

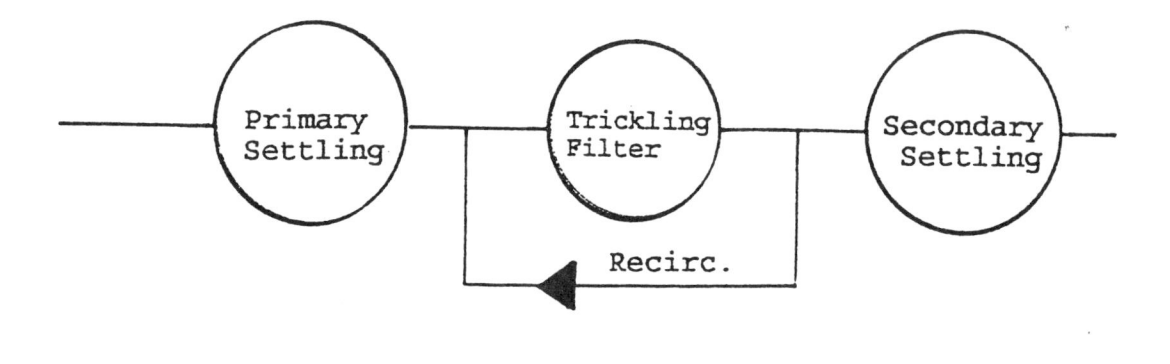

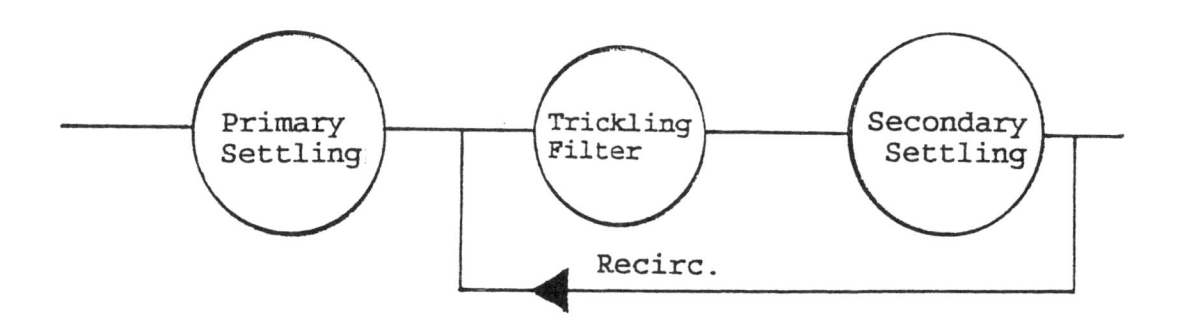

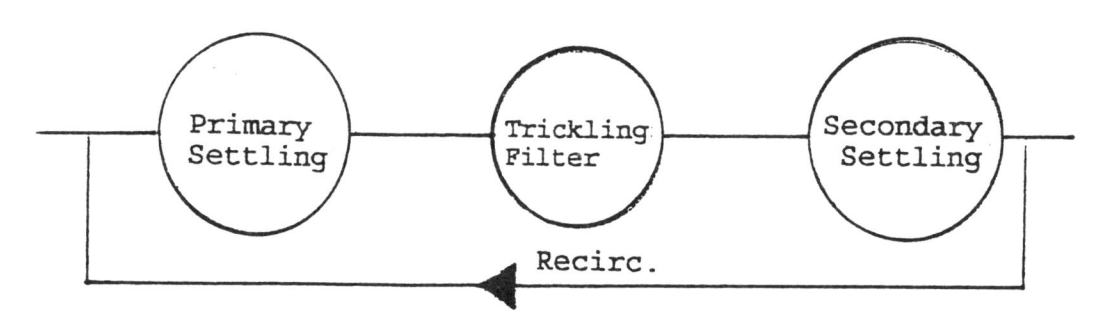

Figure 14.6 Common forms of recirculation.

Maintenance requirements include lubrication of mechanical equipment, removal of debris from the surface and orifices, as well as adjustment of flow patterns and maintenance associated with the settling tank.

Expected performance ranges vary for each classification of trickling filter. Moreover, the levels of BOD_5 and suspended solids removal are dependent on the type of filter.

14.4.4 PROCESS CALCULATIONS

There are several calculations that are useful in the operation of a trickling filter. These include total flow, hydraulic loading, and organic loading.

14.4.4.1 Total Flow

If the recirculated flow rate is given, total flow is

$$\text{Total Flow, MGD} = \text{Influent Flow, MGD} + \text{Recirculation Flow, MGD} \quad (14.9)$$

$$\text{Total Flow, gpd} = \text{Total Flow, MGD} \times 1{,}000{,}000 \, \text{gal}/\text{MG}$$

✓ The total flow to the trickling filter includes the influent flow and the recirculated flow. This can be determined using the recirculation ratio.

$$\text{Total Flow, MGD} = \text{Influent Flow} \times (\text{Recirculation Rate} + 1.0)$$

Example 14.3

Problem:

The trickling filter is currently operating with a recirculation rate of 1.5. What is the total flow applied to the filter when the influent flow rate is 3.65 MGD?

Solution:

$$\text{Total Flow, MGD} = 3.65 \, \text{MGD} \times (1.5 + 1.0)$$

$$= 9.13 \, \text{MGD}$$

14.4.4.2 Hydraulic Loading

Calculating the hydraulic loading rate is important in accounting for both the primary effluent as well as the recirculated trickling filter effluent. Both of these are combined before being applied to the surface of the filter. The hydraulic loading rate is calculated based on the surface area of the filter.

Example 14.4

Problem:

A trickling filter 90 ft in diameter is operated with a primary effluent of 0.488 MGD and a recirculated effluent flow rate of 0.566 MGD. Calculate the hydraulic loading rate on the filter in units gpd/ft^2.

Solution:

The primary effluent and recirculated trickling filter effluent are applied together across the surface of the filter, therefore

$$0.488 \, \text{MGD} + 0.566 \, \text{MGD} = 1.054 \, \text{MGD} = 1{,}054{,}000 \, \text{gpd}$$

$$\text{Circular surface area} = 0.785 \times (\text{diameter})^2$$

$$= 0.785 \times (90 \, \text{ft})^2$$

$$= 6{,}359 \, \text{ft}^2$$

$$\frac{1{,}054{,}000 \, \text{gpd}}{6{,}359 \, \text{ft}^2} = 165.7 \, \text{gpd}/\text{ft}^2$$

✓ Hydraulic loading is to be discussed in greater detail in Volume 2 (Intermediate Level).

14.4.4.3 Organic Loading Rate

As mentioned earlier, trickling filters are sometimes classified by the organic loading rate applied. The organic loading rate is expressed as a certain amount of BOD applied to a certain volume of media.

Example 14.5

Problem:

A trickling filter 50 ft in diameter receives a primary effluent flow rate of 0.445 MGD. Calculate the organic loading rate in units of pounds of BOD applied per day per 1,000 ft^3 of media volume. The primary effluent BOD concentration is 85 mg/L. The media depth is 9 ft.

Solution:

$$0.445 \text{ MGD} \times 85 \text{ mg/L} \times 8.34 \text{ lb/gal} = 315.5 \text{ lb BOD applied/day}$$

$$\text{Surface Area} = 0.785 \times (50)^2$$

$$= 1962.5 \text{ ft}^2$$

$$\text{Area} \times \text{Depth} = \text{Volume}$$

$$1962.5 \text{ ft}^2 \times 9 \text{ ft} = 17{,}662.5 \text{ (TF Volume)}$$

✓ *Note:* To determine the pounds of BOD per 1000 ft^3 in a volume of thousands of cubic feet, we must set up the equation as shown below.

$$\frac{315.5 \text{ lb BOD/day}}{17{,}662.5 \text{ ft}^3} \times \frac{1{,}000}{1{,}000}$$

Regrouping the numbers and the units together:

$$= \frac{315.5 \times 1{,}000}{17{,}662.5 \text{ ft}^3} \times \frac{\text{lb BOD/day}}{1{,}000 \text{ ft}^3}$$

$$= 17.9 \frac{\text{lb BOD/day}}{1{,}000 \text{ ft}^3}$$

✓ *Note:* Organic loading is discussed in greater detail in Volume 2 (Intermediate Level).

14.4.4.4 Settling Tank

In the operation of settling tanks that follow trickling filters, various calculations are routinely made to determine detention time, surface settling rate, hydraulic loading, and sludge pumping.

✓ These calculations are covered in greater detail in Volumes 2 and 3.

14.5 ROTATING BIOLOGICAL CONTACTORS

The rotating biological contactor (RBC) is a biological treatment system (see Figure 14.7) and is a variation of the attached growth idea provided by the trickling filter. Still relying on microorganisms that grow on the surface of a medium, the RBC is instead a fixed film biological treatment device—the basic biological process, however, is similar to that occurring in the trickling filter. An RBC consists of a series of closely spaced (mounted side by side), circular, plastic (synthetic) disks that are typically about 11.5 ft in diameter and are attached to a rotating horizontal shaft (see Figure 14.8). Approximately 40% of each disk is submersed in a tank containing the wastewater to be treated. As the RBC rotates, the attached biomass film (zoogleal slime) that grows on the surface of the disks moves into and out of the wastewater. While submerged in the wastewater, the microorganisms absorb organics; while they are rotated out of the wastewater, they are supplied with needed oxygen for aerobic decomposition. As the zoogleal slime reenters the wastewater, excess solids and waste products are stripped off the media as sloughings. These sloughings are transported with the wastewater flow to a settling tank for removal.

Modular RBC units are placed in series (see Figure 14.7)—simply because a single contactor is not sufficient to achieve the desired level of treatment; the resulting treatment achieved exceeds conventional secondary treatment. Each individual contactor is called a stage, and the group is known as a train. Most RBC systems consist of two or more trains with three or more stages in each. The key advantage in using RBCs instead of trickling filters is that RBCs are easier to operate under varying load conditions, because it is easier to keep the solid medium wet at all times. Moreover, the level of nitrification that can be achieved by an RBC system is significant—especially when multiple stages are employed.

14.5.1 RBC EQUIPMENT

The equipment that makes up an RBC includes the rotating biological contactor (the media: either standard or high density), a center shaft, drive system, tank, baffles, housing or cover, and a settling tank.

The rotating biological contactor consists of circular sheets of synthetic material (usually plastic) that are mounted side by side on a shaft. The sheets (media) contain large amounts of surface area for growth of the biomass.

The center shaft provides the support for the disks of media and must be strong enough to support the weight of the media and the biomass. Experience has indicated that a major problem has been the collapse of the support shaft.

The drive system provides the motive force to rotate the disks and shaft. The drive system may be

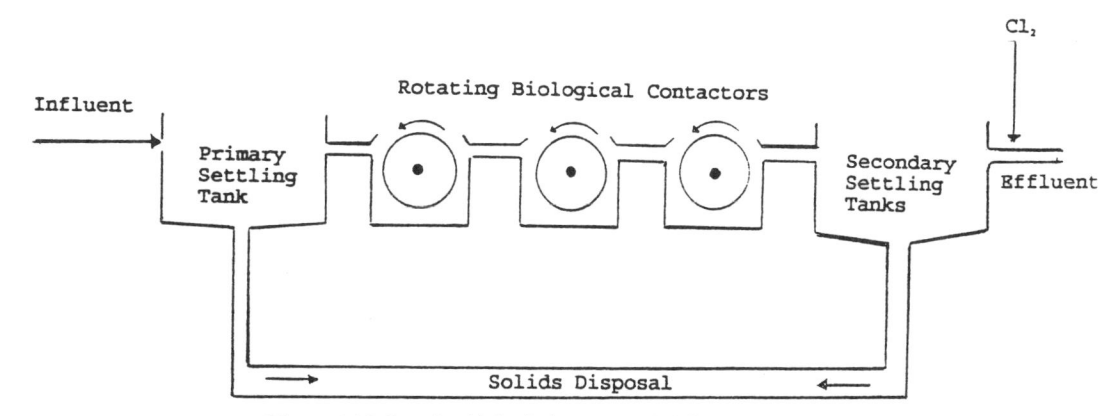

Figure 14.7 Rotating biological contactor (RBC) treatment system.

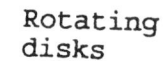

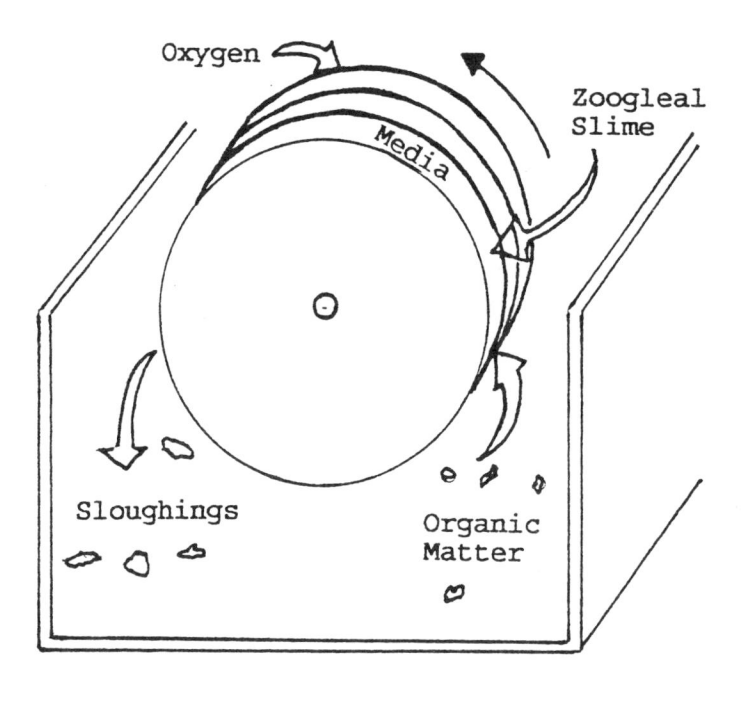

Figure 14.8 Rotating biological contactor (RBC) cross section and treatment system.

mechanical, air driven, or a combination of both. When the drive system does not provide uniform movement of the RBC, major operational problems can arise.

The tank holds the wastewater that the RBC rotates in. It should be large enough to permit variation of the liquid depth and detention time.

Baffles are required to permit proper adjustment of the loading applied to each stage of the RBC process. Adjustment can be made to increase or decrease the submergence of the RBC.

RBC stages are normally enclosed in some type of protective structure (cover) to prevent loss of biomass due to severe weather changes (snow, rain, temperature, wind, sunlight, etc.). In many instances, this housing greatly restricts access to the RBC.

The settling tank is provided to remove the sloughing material created by the biological activity and is similar in design to the primary settling tank. The settling tank provides 2- to 4-hour detention time to permit settling of lighter biological solids.

14.5.2 RBC OPERATION

During normal operation, operator vigilance is required to observe the RBC movement, slime color, and appearance. However, if the unit is covered, observations may be limited to that portion of the media that can be viewed through the access door. Slime color and appearance can indicate process condition. For example, when slime color and/or appearance is

- gray, shaggy slime growth—indicates normal operation

- reddish brown, golden shaggy growth—indicates nitrification
- white chalky appearance—indicates high sulfur concentrations
- no slime—indicates severe temperature or pH changes

Sampling and testing should be conducted daily for dissolved oxygen content and pH. BOD_5 and suspended solids testing should also be accomplished to aid in assessing performance.

14.5.3 RBC: EXPECTED PERFORMANCE

The RBC normally produces a high-quality effluent:

BOD_5	85–95%
Suspended solids removal	85–95%

The RBC treatment process may also significantly reduce (if designed for this purpose) the levels of organic nitrogen and ammonia nitrogen.

14.5.4 RBC: PROCESS CONTROL CALCULATIONS

There are several process control calculations that may be useful in the operation of an RBC. These include hydraulic loading, soluble BOD, and organic loading. Settling tank calculations and sludge pumping calculations may be helpful for evaluation and control of the settling tank following the RBC.

14.5.4.1 RBC: Hydraulic Loading Rate

The RBC media surface area is normally specified by the manufacturer, and the hydraulic loading rate is based on the media surface area, usually in square feet (ft^2). Hydraulic loading on an RBC can range from 1 to 3 gpd/ft^2.

Let's look at an example problem.

Example 14.6

Problem:

An RBC treats a primary effluent flow rate of 0.233 MGD. What is the hydraulic loading rate in gpd/ft^2 if the media surface area is 96,600 ft^2?

Solution:

$$\frac{233{,}000 \text{ gpd}}{96{,}600 \text{ ft}^2} = 2.41 \text{ gpd}/\text{ft}^2$$

14.5.4.2 RBC: Organic Loading Rate

If the soluble BOD concentration is known, the organic loading on an RBC can be determined. Organic loading on an RBC based on soluble BOD concentration can range from 3 to 4 lb/d/1,000 ft^2.

Example 14.7

Problem:

An RBC has a total media surface area of 102,500 ft² and receives a primary effluent flow rate of 0.269 MGD. If the soluble BOD concentration of the RBC influent is 159 mg/L, what is the organic loading rate in lb/1,000 ft²?

Solution:

$$0.269 \text{ MGD} \times 159 \text{ mg/L} \times \frac{8.34 \text{ lb}}{1 \text{ gal}} = 356.7 \text{ lb/d}$$

$$\frac{356.7 \text{ lb/d}}{102,500 \text{ ft}^2} \times \frac{1,000 \text{ (number)}}{1,000 \text{ (unit)}} = 3.48 \text{ lb/d/1,000 ft}^2$$

14.6 CHAPTER REVIEW QUESTIONS

14-1 What type of waste treatment pond is most common?

14-2 Give three classifications of ponds based upon their location in the treatment system.

14-3 Describe the processes occurring in a raw sewage stabilization pond (facultative).

14-4 How do the changes in the season affect the quality of the discharge from a stabilization pond?

14-5 A wastewater treatment pond has an average length of 690 ft with an average width of 425 ft. If the flow rate to the pond is 300,000 gal each day and is operated at a depth of 6 ft, what is the hydraulic detention time in days?

14-6 A pond 730 ft long and 410 ft wide receives an influent flow rate of 0.66 ac-ft/d. What is the hydraulic loading rate on the pond in inches per day?

14-7 Name three main parts of the trickling filter, and give the purpose or purposes of each part.

14-8 Name three categories of trickling filter based on their organic loading rate.

14-9 Which classification of trickling filter produces the highest quality effluent?

14-10 What is the purpose of recirculation?

14-11 The recirculation rate is 0.80. The influent flow rate is 2.3 MGD. What is the total flow being applied to the filter in MGD?

14-12 Why is a settling tank required following the trickling filter?

14-13 List three things that should be checked as part of the normal operations and maintenance procedures for a trickling filter.

14-14 A trickling filter 90 ft in diameter treats a primary effluent flow rate of 0.288 MGD. If the recirculated flow to the clarifier is 0.366 MGD, what is the hydraulic loading rate on the trickling filter in gallons per day per square foot (gpd/ft^2)?

14-15 A treatment plant receives a flow rate of 3.0 MGD. If the trickling filter effluent is recirculated at a rate of 4.30 MGD, what is the recirculation ratio?

14-16 Describe the RBC.

14-17 Describe the process occurring in the RBC process.

14-18 Can an RBC be operated without primary settling?

14-19 What does chalky white biomass indicate?

14-20 Name two types of RBC media.

14-21 What makes the RBC similar to the trickling filter?

14-22 What makes the RBC perform at approximately the same levels of performance throughout the year?

14-23 Describe the appearance of the slime when the RBC is operating properly. What happens if the RBC is exposed to a wastewater containing high amounts of sulfur?

14-24 The slime in the first stages of the RBC is gray and shaggy. The slime in the last two stages of the train are reddish brown. What does this indicate?

14-25 An RBC unit treats a flow rate of 0.45 MGD. The two shafts used provide a total surface area of 200,000 ft^2. What is the hydraulic loading on the unit in gpd/ft^2?

CHAPTER 15

Activated Sludge

15.1 INTRODUCTION

THE biological treatment systems discussed to this point [ponds, trickling filters, and rotating biological contactors (RBCs)] have been around for years. The trickling filter, for example, has been around and successfully used since the late 1800s. The problem with ponds, trickling filters, and RBCs is that they are temperature sensitive, remove less BOD, and trickling filters, for example, cost more to build than the activated sludge systems that were later developed.

✓ *Note:* Although trickling filters and other systems cost more to build than activated sludge systems, it is important to point out that activated sludge systems cost more to operate because of the need for energy to run pumps and blowers.

15.2 ACTIVATED SLUDGE PROCESS: OPERATION OF

The activated sludge process is designed to remove BOD_5 and suspended matter through aerobic decomposition. Nitrogen and phosphorus may also be removed if process controls are properly adjusted.

✓ Removal of nutrients may require inclusion of anaerobic and/or anoxic stages.

Though the example wastewater treatment system flow diagram shown in Figure 15.1 illustrates the activated sludge process, it should be pointed out that activated sludge processes may or may not follow primary treatment. The need for primary treatment is determined by the process modification selected for use. However, all activated sludge systems include a settling tank following the aeration basin. For our purposes, we use Figure 15.1 to aid in explaining the process. As indicated in Figure 15.1, the key biological unit in the process is the aeration tank, which receives effluent from the primary clarifier. It also receives a mass of recycled biological organisms from the secondary settling tank; this is known as *activated sludge*. To maintain aerobic conditions, air or oxygen is pumped into the tank via blowers, and the mixture is kept thoroughly agitated by mixers. After about 6 to 8 hours of agitation, the wastewater (mixed liquor) flows into the secondary settling tank where the solids, mostly bacterial masses, are separated from the liquid by subsidence. A portion of those solids is returned to the aeration tank to maintain the proper bacterial population there, while the remainder must be processed and disposed of.

There are a number of factors that affect the performance of an activated sludge system. These are listed as follows:

- temperature
- return rates

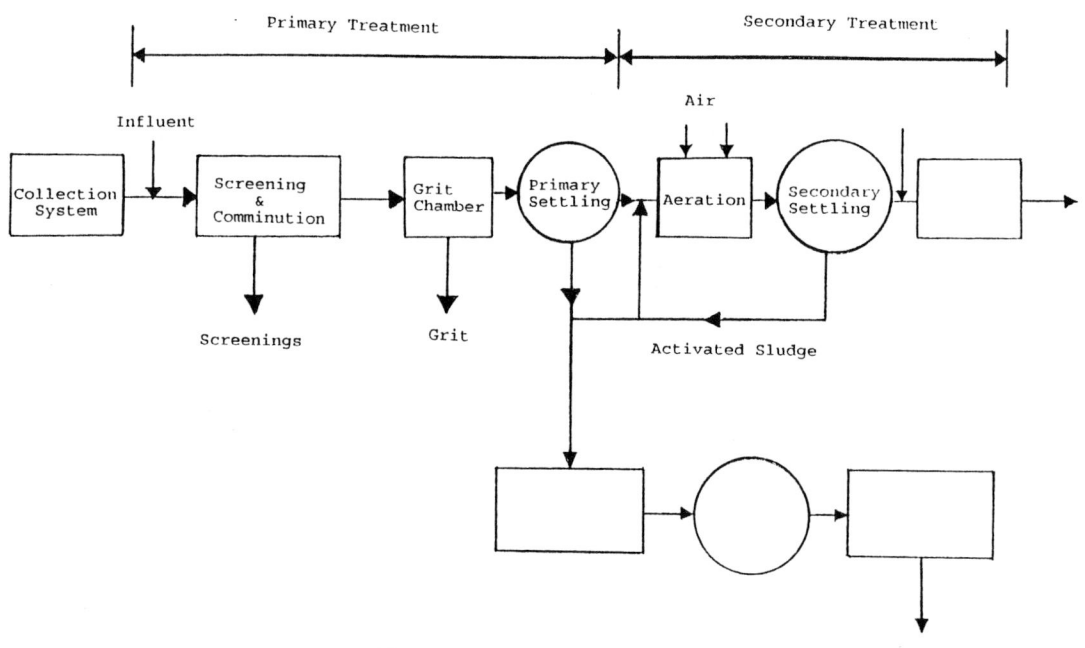

Figure 15.1 Activated sludge process.

- amount of oxygen available
- amount of organic matter available
- pH
- waste rates
- aeration time
- wastewater toxicity

To obtain the desired level of performance in an activated sludge system, a proper balance must be maintained between the amount of food (organic matter), organisms (activated sludge), and oxygen (dissolved oxygen, DO). The majority of problems with the activated sludge process result from an imbalance between these three items.

15.3 ACTIVATED SLUDGE PROCESS: EQUIPMENT

The equipment requirements for the activated sludge process are more complex than other processes discussed. Equipment includes an aeration tank, aeration, system settling tank, return sludge, and waste sludge. These are discussed in the following.

15.3.1 AERATION TANK

The aeration tank is designed to provide the required detention time (depends on the specific modification) and to ensure that the activated sludge and the influent wastewater are thoroughly mixed. Tank design normally attempts to ensure no dead spots are created.

15.3.2 AERATION

Aeration can be mechanical or diffused. Mechanical aeration systems use agitators or mixers to mix air and mixed liquor. Some systems use sparge rings to release air directly into the mixer.
Diffused aeration systems use pressurized air released through diffusers near the bottom of the

tank. Efficiency is directly related to the size of the air bubbles produced. Fine bubble systems have a higher efficiency. The diffused air system has a blower to produce large volumes of low-pressure air (5 to 10 psi), air lines to carry the air to the aeration tank, and headers to distribute the air to the diffusers, which release the air into the wastewater.

15.3.3 SETTLING TANK

Activated sludge systems are equipped with plain settling tanks designed to provide 2 to 4 hours of hydraulic detention time.

15.3.4 RETURN SLUDGE

The return sludge system includes pumps, a timer or variable speed drive to regulate pump delivery, and a flow measurement device to determine actual flow rates.

15.3.5 WASTE SLUDGE

In some cases, the waste-activated sludge withdrawal is accomplished by adjusting valves on the return system. When a separate system is used, it includes pump(s), timer or variable speed drive, and a flow measurement device.

15.4 ACTIVATED SLUDGE PROCESS: MODIFICATIONS

Many activated sludge process modifications exist. Each modification is designed to address specific conditions or problems. The modifications discussed in this handbook are

- conventional activated sludge
- contact stabilization activated sludge
- extended aeration activated sludge
- oxidation ditch activated sludge

15.4.1 CONVENTIONAL ACTIVATED SLUDGE

- When the conventional activated-sludge modification is employed, primary treatment is required.
- Excellent treatment is provided; however, large aeration tank capacity is required, and construction costs are high.
- In operation, initial oxygen demand is high. Moreover, the process is very sensitive to operational problems (e.g., bulking).

15.4.2 CONTACT STABILIZATION

- Contact stabilization does not require primary treatment.
- During operation, organisms collect organic matter (during contact).
- Solids and activated sludge are separated from flow via settling.
- Activated sludge and solids are aerated for 3 to 6 hours (stabilization).

✓ *Note:* Return sludge is aerated before it is mixed with influent flow.

- The activated sludge oxidizes available organic matter.

- While the process is complicated to control, it requires less tank volume than other modifications and can be prefabricated as a package unit for flows of 0.05 to 1.0 MGD.
- A disadvantage is that common process control calculations do not provide usable information.

15.4.3 EXTENDED AERATION

- does not require primary treatment
- is used frequently for small flows, such as schools and subdivisions
- uses 24-hour aeration
- produces low BOD_5 effluent
- produces the least amount of waste-activated sludge
- is capable of achieving 95% or more removals of BOD_5
- can produce effluent low in organic and ammonia nitrogen

15.4.4 OXIDATION DITCH

- does not require primary treatment
- process is similar to the extended aeration process

Table 15.1 lists the process parameters for each of the four activated-sludge modifications.

15.5 AERATION TANK OBSERVATIONS

Wastewater operators are required to monitor or to make certain observations of treatment unit processes to ensure optimum performance and to make adjustments when required. In monitoring the operation of an aeration tank, there are three physical parameters (turbulence, surface foam and scum, and sludge color and odor) that the operator should look for that aid in determining how the process is operating and indicate if any operational adjustments need to be made. Let's take a look at a few of these parameters.

15.5.1 TURBULENCE

Normal operation of an aeration basin includes a certain amount of turbulence. This turbulent

TABLE 15.1. Activated-Sludge Modifications.

Parameter	Conventional	Contact Stabilization	Extended Aeration	Oxidation Ditch
Aeration Time, h	4–8	0.5–1.5 (contact) 3–6 (reaeration)	24	24
Settling Time, h	2–4	2–4	2–4	2–4
Return Rate, % of Influent Flow	25–100	25–100	25–100	25–100
MLSS, mg/L	1,500–4,000	1,000–3,000 3,000–8,000	2,000–6,000	2,000–6,000
DO, mg/L	1–3	1–3	1–3	1–3
SSV_{30}, mL/L	400–700	400–700 (contact)	400–700	400–700
Food:Mass Ratio lb BOD_5/lb MLVSS	0.2–0.5	0.2–0.6 (contact)	0.05–0.15	0.05–0.15
MCRT (Whole System), Days	5–15	N/A	20–30	20–30
% Removal BOD_5	85–95%	85–95%	85–95%	85–95%
% Removal TSS	85–95%	85–95%	85–95%	85–95%
Primary Treatment	Yes	No	No	No

action is, of course, required to ensure a consistent mixing pattern. However, whenever excessive, deficient or non-uniform mixing occurs, adjustments may be necessary to air flow—or diffusers may need cleaning or replacement.

15.5.2 SURFACE FOAM AND SCUM

The type, color, and amount of foam or scum present may indicate the required wasting strategy to be employed. Types of foam include the following:

- fresh, crisp, white foam—Moderate amounts of a crisp white foam are usually associated with activated sludge processes that are producing an excellent final effluent. Adjustment: None, normal operation.
- thick, greasy, dark tan foam—A thick greasy dark tan or brown foam or scum normally indicates an old sludge that is over-oxidized. Adjustment: Indicates old sludge, more wasting required.
- white billowing foam—Large amounts of a white, soap suds-like foam indicate a very young under-oxidized sludge. Adjustment: Young sludge, less wasting required.

15.5.3 SLUDGE COLOR AND ODOR

Though not as reliable an indicator of process operations as foam, sludge color and odor are useful indicators. Colors and odors that are important include:

- chocolate brown, earthy odor—indicates normal operation
- light tan or brown/no odor—indicates sand and clay from infiltration/inflow. Adjustment: extremely young sludge, decrease wasting
- dark brown/earthy odor—indicates old sludge, high solids. Adjustment: increase wasting
- black color/rotten egg odor—indicates septic conditions. Adjustment: increase aeration

15.6 FINAL SETTLING TANK (CLARIFIER) OBSERVATIONS

Settling tank observations include flow pattern (normally uniform distribution), settling, amount and type of solids leaving with the process effluent (normally very low), and the clarity or turbidity of the process effluent (normally very clear).

Observations should include the following conditions:

(1) Sludge bulking—occurs when solids are evenly distributed throughout the tank and leaving over the weir in large quantities.
(2) Sludge solids washout—sludge blanket is down but solids are flowing over the effluent weir in large quantities. Control tests indicate a good quality sludge.
(3) Clumping—large "clumps" or masses of sludge (several inches or more) rise to the top of the settling tank.
(4) Ashing—fine particles of gray to white material flowing over the effluent weir in large quantities.
(5) Straggler floc—small, almost transparent, very fluffy, buoyant solids particles (1/8" to 1/4" diameter rising to the surface). Usually is accompanied by a very clean effluent. Usually new growth, most noted in the early morning hours. Sludge age is slightly below optimum.
(6) Pin floc—very fine solids particles (usually less than 1/2" in diameter) suspended throughout lightly turbid liquid. Usually the result of an over-oxidized sludge.

15.7 PROCESS CONTROL TESTING AND SAMPLING

Process control testing may include settleability testing to determine the settled sludge volume; suspended solids testing to determine influent and mixed liquor suspended solids; return activated sludge solids and waste activated sludge concentrations; determination of the volatile content of the mixed liquor suspended solids; dissolved oxygen and pH of the aeration tank; BOD_5 and/or chemical oxygen demand (COD) of the aeration tank influent and process effluent; and microscopic evaluation of the activated sludge to determine the predominant organism.

The following sections describe most of the common process control tests.

15.7.1 pH

pH is tested daily with a sample taken from the aeration tank influent and process effluent. pH is normally close to 7.0 (normal) with the best pH range from 6.5 to 8.5 (however, a pH range of 6.5 to 9.0 is satisfactory). Keep in mind that the effluent pH may be lower because of nitrification.

15.7.2 TEMPERATURE

Temperature is important because it forecasts the following:

Temperature Increases: Organism activity increases
Aeration efficiency decreases
Oxygen solubility decreases
Temperature Decreases: Organism activity decreases
Aeration efficiency increases
Oxygen solubility increases

15.7.3 DISSOLVED OXYGEN

Content of dissolved oxygen (DO) in the aeration process is critical to performance. DO should be tested at least daily (peak demand). Optimum is determined for individual plants, but normal is from 1 to 3 mg/L. If the system contains too little DO, the process will become septic. If it contains too much DO, energy and money are wasted.

15.7.4 SETTLED SLUDGE VOLUME (SETTLEABILITY)

Settled sludge volume (SSV) is determined at specified times during sample testing. Thirty- and 60-minute observations are used for control. Subscripts (SSV_{30} and SSV_{60}) indicate settling time. The test is performed on aeration tank effluent samples.

$$SSV = \frac{\text{Milliliters of Settled Sludge 1,000 mL/L}}{\text{Milliliters of Sample}} \quad (15.1)$$

$$\%SSV = \frac{\text{Milliliters of Settled Sludge} \times 100}{\text{Milliliters of Sample}} \quad (15.2)$$

Under normal conditions, sludge settles as a mass, producing clear supernatant with SSV_{60} in the range of 400–700 mL/L. When higher values are indicated, this may indicate excessive solids (old sludge) and/or bulking conditions. Rising solids (if sludge is well oxidized) may rise after 2 or more hours. However, rising solids in less than 1 hour indicates a problem.

✓ Running the settleability test with a diluted sample can assist in determining if the activated sludge is old (too many solids) or bulking (not settling). Old sludge will settle to a more compact level when diluted.

15.7.5 CENTRIFUGE TESTING

The centrifuge test provides a quick, relatively easy control test for the solids level in the aerator, but does not usually correlate with mixed liquor suspended solids (MLSS) results. Results are directly affected by variations in sludge quality.

15.7.6 MICROSCOPIC EXAMINATION

The activated sludge process cannot operate as designed without the presence of microorganisms. Thus, microscopic examination of an aeration basin sample, to determine the presence and the type of microorganisms, is important. Different species prefer different conditions; therefore, the presence of different species can indicate process condition.

✓ *Note:* It is important to point out that during microscopic examination, identifying of all organisms present is not required, but identification of the predominant species is required.

Table 15.2 lists process conditions indicated by the presence and population of certain microorganisms.

15.7.7 SLUDGE BLANKET DEPTH

Sludge blanket depth refers to the distance from the surface of the liquid to the solids-liquid interface or the thickness of the sludge blanket as measured from the bottom of the tank to the

TABLE 15.2. Process Condition Versus Organisms Present/Population.

Process Condition	Organism Population
Poor BOD$_5$, and TSS Removal Mainly dispersed bacteria No floc formation Very cloudy effluent	Predominance of amoeba and flagellates A few ciliates present
Poor Quality Effluent Dispersed bacteria Some floc formation Cloudy effluent	Predominance of amoeba and flagellates Some free-swimming ciliates
Satisfactory Effluent Good floc formation Good settleability Good clarity	Predominance of free-swimming ciliates Few amoeba and flagellates
High Quality Effluent Excellent floc formation Excellent settleability High effluent clarity	Predominance of stalked ciliates Some free-swimming ciliates A few rotifers A few flagellates
Effluent High TSS and Low BOD$_5$ High settled sludge volume Cloudy effluent	Predominance of rotifers Large numbers of stalked ciliates A few free-swimming ciliates No flagellates

solids-liquid interface. Part of the operator's sampling routine, this measurement is taken directly in the final clarifier. Sludge blanket depth is dependent upon hydraulic load, return rate, clarifier design, waste rate, sludge characteristics, and temperature. If all other factors remain constant, the blanket depth will vary with the amount of solids in the system and the return rate; thus, it will vary throughout the day.

15.7.8 SUSPENDED SOLIDS AND VOLATILE SUSPENDED SOLIDS

Suspended solids and volatile suspended solids concentrations of the mixed liquor (MLSS), the return activated sludge (RAS), and waste activated sludge (WAS) are routinely sampled and tested because they are critical to process control.

15.8 PROCESS CONTROLS

In the routine performance of their duties, wastewater operators make process control adjustments to various unit processes, including the activated sludge process. In the following, a summary is provided of the process controls available for the activated sludge process and the result that will occur from adjustment of each.

15.8.1 PROCESS CONTROL: RETURN RATE

(1) Condition: *Return rate too high*
    Result:
    - hydraulic overloading of aeration and settling tanks
    - reduced aeration time
    - reduced settling time
    - loss of solids over time
(2) Condition: *Return rate too low*
    Result:
    - septic return
    - solids buildup in settling tank
    - reduced MLSS in aeration tank
    - loss of solids over weir

15.8.2 PROCESS CONTROL: WASTE RATE

(1) Condition: *Waste rate too high*
    Result:
    - reduced MLSS
    - decreased sludge density
    - increased sludge volume index (SVI)
    - decreased mean cell residence time (MCRT)
    - increased F:M ratio
(2) Condition: *Waste rate too low*
    Result:
    - increased MLSS
    - increased sludge density
    - decreased SVI

- increased MCRT
- decreased F:M ratio

15.8.3 PROCESS CONTROL: AERATION RATE

(1) Condition: *Aeration rate too high*
Result:
- wasted energy
- increased operating cost
- rising solids
- breakup of activated sludge

(2) Condition: *Aeration rate too low*
Result:
- septic aeration tank
- poor performance
- loss of nitrification

15.9 TROUBLESHOOTING OPERATIONAL PROBLEMS

Without a doubt, the most important dual function performed by the wastewater operator is the identification of process control problems and the implementation of the appropriate actions to correct the problem(s). In this section, typical aeration system operational problems are listed with their symptoms, causes, and the appropriate corrective actions required to restore the unit process to a normal or optimal performance level.

15.9.1 SYMPTOM 1

The solids blanket is flowing over the effluent weir (classic bulking). Settleability test shows no settling.

(1) *Cause:* Organic overloading
   Corrective action: Reduce organic loading
(2) *Cause:* Low pH
   Corrective action: Add alkalinity
(3) *Cause:* Filamentous growth
   Corrective action: Add nutrients; add chlorine or peroxide to return
(4) *Cause:* Nutrient deficiency
   Corrective action: Add nutrients
(5) *Cause:* Toxicity
   Corrective action: Identify source; implement pretreatment
(6) *Cause:* Overaeration
   Corrective action: Reduce aeration during low flow periods

15.9.2 SYMPTOM 2

Solids settled properly in settleability test but large amounts of solids were lost over effluent weir.

(1) *Cause:* Billowing solids due to short circuiting
   Corrective action: Identify short circuiting cause and eliminate if possible

15.9.3 SYMPTOM 3

Large amounts of small pin-head-sized solids leaving settling tank.

(1) *Cause:* Old sludge
   Corrective action: Reduce sludge age (gradual change is best); increase waste rate
(2) *Cause:* Excessive turbulence
   Corrective action: Decrease turbulence (adjust aeration during low flows)

15.9.4 SYMPTOM 4

Large amount of light floc (low BOD_5 and high solids) leaving settling tank.

(1) *Cause:* Extremely old sludge
   Corrective action: Reduce age; increase waste

15.9.5 SYMPTOM 5

Large amount of small translucent particles (1/16–1/8″) are leaving the settling tank.

(1) *Cause:* Rapid solids growth
   Corrective action: Increase sludge age
(2) *Cause:* Slightly young activated sludge
   Corrective action: Decrease waste

15.9.6 SYMPTOM 6

Solids settling properly but rise to surface within a short time. Many small (1/4″) to large (several feet) clumps of solids on surface of settling tank.

(1) *Cause:* Denitrification
   Corrective action: Increase rate of return; adjust sludge age to eliminate nitrification
(2) *Cause:* Overaeration
   Corrective action: Reduce aeration

15.9.7 SYMPTOM 7

Return activated sludge has a rotten egg odor.

(1) *Cause:* Return is septic
   Corrective action: Increase aeration rate
(2) *Cause:* Return rate is too low
   Corrective action: Increase rate of return

15.9.8 SYMPTOM 8

Activated sludge organisms die during a short time.

(1) *Cause:* Influent contained toxic material
   Corrective action: Isolate activated sludge (if possible); return all available solids; stop wasting; increase return rate; implement pretreatment program

15.9.9 SYMPTOM 9

Surface of aeration tank covered with thick, greasy foam.

(1) *Cause:* Extremely old activated sludge
   Corrective action: Reduce activated sludge age; increase wasting; use foam control sprays
(2) *Cause:* Excessive grease and oil in system
   Corrective action: Improve grease removal; use foam control sprays; implement pretreatment program
(3) *Cause:* Froth-forming bacteria
   Corrective action: Remove froth-forming bacteria

15.9.10 SYMPTOM 10

Large "clouds" of billowing white foam on the surface of the aeration tank.

(1) *Cause:* Young activated sludge
   Corrective action: Increase sludge age; decrease wasting; use foam control sprays
(2) *Cause:* Low solids in aeration tank
   Corrective action: Increase sludge age; decrease wasting; use foam control sprays
(3) *Cause:* Surfactants (detergents)
   Corrective action: Eliminate surfactants; use foam control sprays; add antifoam

15.10 PROCESS CONTROL CALCULATIONS

As with other wastewater treatment unit processes, process control calculations are important tools used by the operator to optimize and control process operations. In this section, we review the most frequently used activated sludge calculations.

15.10.1 SETTLED SLUDGE VOLUME

Settled sludge volume (SSV) is the volume that a settled activated sludge occupies after a specified time. The settling time may be shown as a subscript (i.e., SSV_{60} indicates the reported value was determined at 60 minutes). The settled sludge volume can be determined for any time interval; however, the most common values are the 30-minute reading (SSV_{30}) and 60-minute reading (SSV_{60}). The settled sludge volume can be reported as milliliters of sludge per liter of sample (mL/L) or as a percent settled sludge volume.

$$\text{Settled Sludge Volume, mL/L} = \frac{\text{Settled Sludge Volume, mL}}{\text{Sample Volume, L}} \quad (15.3)$$

✓ 1,000 mL = 1 L

$$\text{Sample Volume, L} = \frac{\text{Sample Volume in milliliters}}{1,000 \text{ mL/L}} \quad (15.4)$$

$$\text{\% Settled Sludge Volume} = \frac{\text{Settled Sludge Volume, mL} \times 100}{\text{Sample Volume, mL}} \quad (15.5)$$

Example 15.1

Problem:

Using the information provided in the table, calculate the SSV_{30} and the % SSV_{60}.

Time	Millileters
Start	2,500
15 minutes	2,250
30 minutes	1,800
45 minutes	1,700
60 minutes	1,600

Solution:

$$\text{Settled Sludge Volume, } (SSV_{30}) = \frac{1,800 \text{ mL}}{2.5 \text{ L}} = 720 \text{ mL/L}$$

$$\text{\% Settled Sludge Volume, } (SSV_{60}) = \frac{1,600 \text{ mL} \times 100}{2,500 \text{ mL}} = 64\%$$

15.10.2 ESTIMATED RETURN RATE

There are many different methods available for estimation of the proper return sludge rate. A simple method described in the *Operation of Wastewater Treatment Plants, Field Study Program* (1986)—developed by the California State University, Sacramento—uses the 60-minute percent settled sludge volume. The $\%SSV_{60}$ can provide an approximation of the appropriate return activated sludge rate. The results of this calculation can then be adjusted based upon sampling and visual observations to develop the optimum return sludge rate.

✓ The $\%SSV_{60}$ must be converted to a decimal percent and total flow rate (wastewater flow and current return rate in million gallons per day must be used).

$$\text{Estimated Return Rate, million gallons per day (MGD)} = \qquad (15.6)$$
$$(\text{Influent Flow, MGD} + \text{Current Return Flow, MGD}) \times \%SSV_{60}$$

Assumes $\%SSV_{60}$ is representative
Assumes return rate, in percent equals $\%SSV_{60}$
Actual return rate is normally set slightly higher

Example 15.2

Problem:

The influent flow rate is 4.2 MGD, and the current return activated sludge flow rate is 1.5 MGD. The SSV_{60} is 38%. Based upon this information, what should be the return sludge rate in million gallons per day (MGD)?

Solution:

$$\text{Return, MGD} = (4.2 \text{ MGD} + 1.5 \text{ MGD}) \times 0.38 = 2.2 \text{ MGD}$$

15.10.3 SLUDGE VOLUME INDEX

Sludge volume index (SVI) is a measure of the settling quality (a quality indicator) of the activated sludge. As the SVI increases, the sludge settles slower, does not compact as well, and is likely to result in an increase in effluent suspended solids. As the SVI decreases, the sludge becomes more dense, settling is more rapid, and the sludge is becoming older. SVI is the volume in milliliters occupied by 1 gram of activated sludge. The settled sludge volume, mL/L and the mixed liquor suspended solids (MLSS) mg/L are required for this calculation.

$$\text{Sludge Volume Index (SVI)} = \frac{\text{SSV, mL/L} \times 1{,}000}{\text{MLSS mg/L}} \quad (15.7)$$

Example 15.3

Problem:

The SSV_{30} is 365 mL/L and the MLSS is 2,365 mg/L. What is the SVI?

Solution:

$$\text{Sludge Volume Index} = \frac{365 \text{ mL/L} \times 1{,}000}{2{,}365 \text{ mg/L}} = 154.3$$

SVI equals 154.3—what does this mean? It means that the system is operating normally with good settling and low effluent turbidity.

How do we know this? We know this because we compare the 154.3 result with the parameters listed below to obtain the expected condition (the result).

SVI Value	Expected Condition (indicates)
Less than 100	Old sludge—possible pin floc
	Effluent turbidity increasing
100–200	Normal operation—good settling
	Low effluent turbidity
Greater than 250	Bulking sludge—poor settling
	High effluent turbidity

15.10.4 WASTE ACTIVATED SLUDGE

The quantity of solids removed from the process as waste activated sludge (WAS) is an important process control parameter that operators need to be familiar with—and, more importantly, know how to calculate.

$$\text{Waste, lb/day} = \text{WAS Conc., mg/L} \times \text{WAS Flow, MGD} \times 8.34 \text{ lb/MG/mg/L} \quad (15.8)$$

Example 15.4

Problem:

The operator wastes 0.44 MGD of activated sludge. The waste activated sludge has a solids concentration of 5,540 mg/L. How many pounds of waste activated sludge are removed from the process?

Solution:

$$\text{Waste, lb/day} = 5{,}540\,\text{mg/L} \times 0.44\,\text{MGD} \times 8.34 = 20{,}329.6\,\text{lb/day}$$

15.10.5 FOOD-TO-MICROORGANISM RATIO (F/M RATIO)

The food-to-microorganism ratio (F/M ratio) is a process control calculation used in many activated sludge facilities to control the balance between available food materials (BOD or COD) and available organisms (mixed liquor volatile suspended solids, MLVSS).

$$\text{F/M Ratio} = \frac{\text{Primary Eff. COD/BOD mg/L} \times \text{Flow MGD} \times 8.34\,\text{lb/mg/L/MG}}{\text{MLVSS mg/L} \times \text{Aerator Volume, MG} \times 8.34\,\text{lb/mg/L/MG}} \qquad (15.9)$$

Typical F/M ratio for activated sludge processes is shown in the following:

Typical F/M Ratio (Activated Sludge Processes)		
Process	lb BOD$_5$ / lb MLVSS	lb COD / lb MLVSS
Conventional	0.2–0.4	0.5–1.0
Contact Stabilization	0.2–0.6	0.5–1.0
Extended Aeration	0.05–0.15	0.2–0.5
Oxidation Ditch	0.05–0.15	0.2–0.5

Example 15.5

Problem:

Given the following data, what is the F/M ratio?

Primary Effluent Flow	2.5 MGD
Primary Effluent BOD	145 mg/L
Primary Effluent TSS	165 mg/L
Effluent Flow	2.2 MGD
Effluent BOD	22 mg/L
Effluent TSS	16 mg/L
Aeration Volume	0.65 MG
Settling Volume	0.30 MG
MLSS	3,650 mg/L
MLVSS	2,550 mg/L
% Waste Volatile	71%
Desired F/M	0.3

Solution:

$$\text{F/M Ratio} = \frac{145\,\text{mg/L} \times 2.5\,\text{MGD} \times 8.34\,\text{lb/mg/L/MG}}{2{,}550\,\text{mg/L} \times 0.65\,\text{MG} \times 8.34\,\text{lb/mg/L/MG}}$$

$$= 0.22\,\text{lb BOD/lb MLVSS}$$

✓ If the MLVSS concentration is not available, it can be calculated if % volatile matter (% VM) of the Mixed Liquor Suspended Solids (MLSS) is known [see Equation (15.10)].

$$\text{MLVSS} = \text{MLSS} \times \% \text{ (decimal) Volatile Matter (VM)} \qquad (15.10)$$

Example 15.6

Problem:

The aeration tank contains 2,985 mg/L of MLSS. Laboratory tests indicate the MLSS is 66% volatile matter. What is the MLVSS concentration in the aeration tank?

Solution:

$$\text{MLVSS, mg/L} = 2{,}985 \text{ mg/L} \times 0.66 = 1{,}970 \text{ mg/L}$$

15.10.6 MEAN CELL RESIDENCE TIME (MCRT)

Mean cell residence time (MCRT) (sometimes called sludge retention time) is a process control calculation used for activated sludge systems. The MCRT calculation illustrated in Example 15.7 uses the entire volume of the activated sludge system (aeration and settling).

Mean Cell Residence Time, days =

$$\frac{[\text{MLSS mg/L} \times (\text{Aeration Vol., MG} + \text{Clarifier Vol., MG}) \times 8.34]}{[(\text{WAS, mg/L} \times \text{WAS flow, MGD} \times 8.34) + (\text{TSS out, mg/L} \times \text{Flow} \times 8.34)]} \qquad (15.11)$$

✓ Due to the length of the MCRT equation, the units for the conversion factor 8.34 have not been included. The dimensions for the 8.34 conversion factor are lb/mg/L/MG.

✓ *Note:* MCRT can be calculated using only the aeration tank solids inventory. When comparing plant operational levels to reference materials, it is important to determine which calculation the reference manual uses to obtain its example values.

Example 15.7

Problem:

Given the following data, what is the MCRT?

Influent Flow	4.2 MGD
Influent BOD	135 mg/L
Influent TSS	150 mg/L
Effluent Flow	4.2 MGD
Effluent BOD	22 mg/L
Effluent TSS	10 mg/L
Aeration Volume	1.20 MG
Settling Volume	0.60 MG
MLSS	3,350 mg/L
Waste Rate	0.080 MGD
Waste Conc.	6,100 mg/L
Desired MCRT	8.5 days

Solution:

$$\text{MRCT} = \frac{[3{,}350 \text{ mg/L} \times (1.2 \text{ MG} + 0.6 \text{ MG}) \times 8.34]}{[(6{,}100 \text{ mg/L} \times 0.08 \text{ MGD} \times 8.34) + (10 \text{ mg/L} \times 4.2 \text{ MGD} \times 8.34)]}$$

$$\text{MCRT} = 11.4 \text{ days}$$

15.11 SOLIDS CONCENTRATION: SECONDARY CLARIFIER

The solids concentration in the secondary clarifier can be assumed to be equal to the solids concentration in the aeration tank effluent. It may also be determined in the laboratory using a core sample taken from the secondary clarifier. The secondary clarifier solids concentration can be calculated as an average of the secondary effluent suspended solids and the return activated sludge suspended solids concentration.

15.12 ACTIVATED SLUDGE PROCESS RECORDKEEPING REQUIREMENTS

Wastewater operators soon learn that recordkeeping is a major requirement and responsibility of their jobs. Records are important (essential) for process control, for providing information on the cause of problems, for providing information for making seasonal changes, and for compliance with regulatory agencies. Records should include sampling and testing data, process control calculations, meter readings, process adjustments, operational problems and corrective action taken, and process observations.

15.13 REFERENCE

Operation of Wastewater Treatment Plants: A Field Study Training Program, Vol. 2, 3rd ed., Sacramento, CA: California State University, 1986.

15.14 CHAPTER REVIEW QUESTIONS

15-1 What two purposes does the air supplied to the aeration basin serve?

15-2 What is the liquid mixture of microorganisms and solids removed from the bottom of the settling tank called?

15-3 What is the mixture of primary effluent and return sludge called?

15-4 What are the three things that must be balanced to make the activated sludge process perform properly?

15-5 What is the major purpose of an activated sludge process?

15-6 Name two types of aeration device used in the activated sludge process.

15-7 List three observations the operator should make as part of the daily operation of the activated sludge process.

15-8 A 2,200 mL sample of activated sludge is allowed to settle for 60 minutes. At the end of the 60 minutes, the sludge has settled 1,300 mL. What is the SSV_{60} of the sample?

15-9 An activated sludge sample has an MLSS concentration of 2,245 mg/L. The SSV_{30} of the sample is 420 mL/L. What is the sludge volume index of the sample?

15-10 The operator wastes 0.069 MGD of activated sludge. The WAS concentration is 8,185 mg/L. How many pounds of activated sludge solid have been removed from the process?

15-11 Which activated sludge process aerates the return sludge before mixing it with the influent flow?

15-12 Name three treatment processes used to reduce sludge volume.

CHAPTER 16

Disinfection: Chlorination/Dechlorination

16.1 INTRODUCTION

AFTER secondary treatment to remove BOD$_5$ and solids, wastewater may still contain large numbers of microorganisms. Some of these microbes may be pathogenic and may cause epidemics if discharged to the receiving waters. Wastewater treatment functions to reduce the possibility of this happening. As with other steps (unit processes) of treatment, there are many processes available to achieve disinfection. The most widely used process is chlorination. Other processes include ultraviolet (UV) light, ozonation (ozone treatment), and bromine chloride additions. Only the chlorination process is discussed in Volume 1 of this handbook; the other disinfection processes are covered in Volumes 2 and 3.

16.2 CHLORINATION

Chlorination for disinfection, as shown in Figure 16.1, follows all other steps of treatment. The purpose of chlorination is to reduce the population of organisms in the wastewater to levels low enough to ensure that pathogenic organisms will not be present in sufficient quantities to cause disease when discharged.

✓ *Note:* You might wonder why it is that chlorination of critical waters, such as natural trout streams, is not normal practice. This practice is strictly prohibited because chlorine and its by-products (i.e., chloramines) are extremely toxic to aquatic organisms.

16.2.1 CHLORINATION TERMINOLOGY

There are several terms used in discussion of disinfection by chlorination. It is important for the operator to be familiar with these terms.

- *Chlorine* a strong oxidizing agent that has strong disinfecting capability. A yellow-green gas that is extremely corrosive and is toxic to humans in extremely low concentrations in air.
- *Contact time* the length of time the disinfecting agent and the wastewater remain in contact.
- *Demand* the chemical reactions that must be satisfied before a residual or excess chemical will appear.
- *Disinfection* refers to the selective destruction of disease-causing organisms. All the organisms are not destroyed during the process. This differentiates disinfection from sterilization, which is the destruction of all organisms.
- *Dose* the amount of chemical being added in milligrams/liter.
- *Feed rate* the amount of chemical being added in pounds per day.

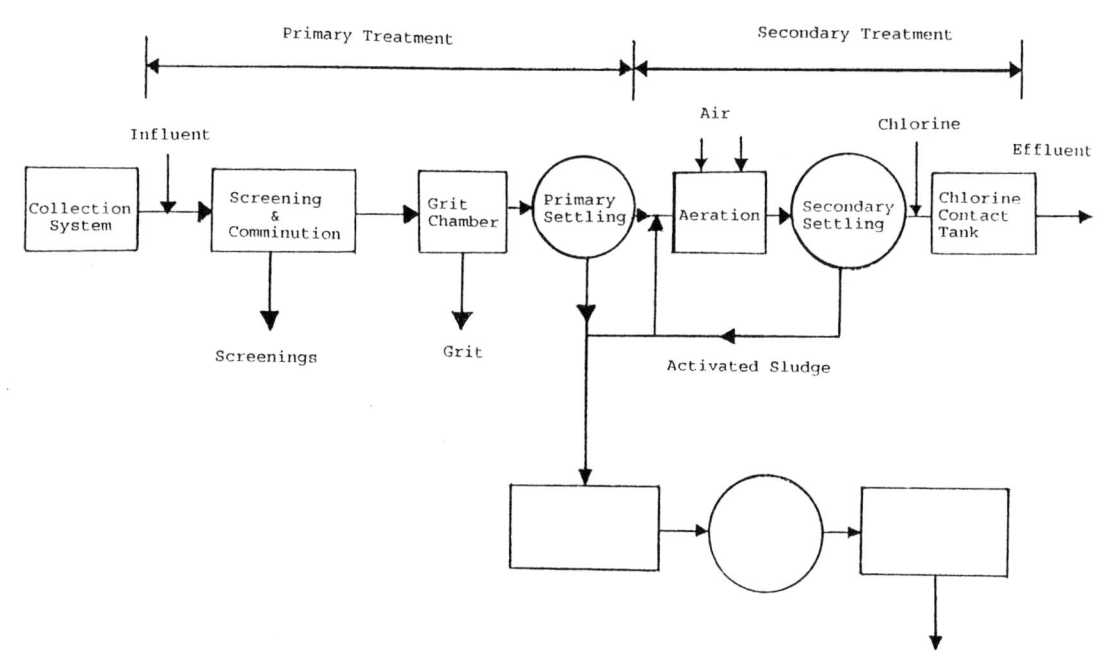

Figure 16.1 Disinfection.

- *Residual* the amount of disinfecting chemical remaining after the demand has been satisfied.
- *Sterilization* the removal of all living organisms.

16.2.2 CHLORINATION: DESCRIPTION

Chlorine is added to wastewater to satisfy all chemical demands (such as sulfide, sulfite, ferrous iron, etc.). When these initial chemical demands have been satisfied, chlorine will react with substances such as ammonia to produce chloramines and other substances that, although not as effective as chlorine, have disinfecting capability. This produces a combined residual that can be measured using residual chlorine test methods. If additional chlorine is added, free residual chlorine can be produced. Due to the chemicals normally found in wastewater, chlorine residuals are normally combined rather than free residuals. Control of the disinfection process is normally based upon maintaining a total residual chlorine of at least 1.0 mg/L for a contact time of at least 30 minutes at design flow.

✓ *Note:* Disinfection is affected by residual level, contact time, and effluent quality. Failure to maintain the desired residual levels for the required contact time will result in lower efficiency and increased probability that disease organisms will be discharged.

16.2.3 CHLORINATION CHEMICALS

Chlorine used in the disinfection process normally is in the form of hypochlorite (similar to that used for home swimming pools) or free chlorine gas.

16.2.3.1 Hypochlorite

Hypochlorite, though there are some minor hazards associated with its use (skin irritation, nose irritation, and burning eyes), is relatively safe to work with. It is normally available in dry form as a white powder, pellet, or tablet or in liquid form. It can be added directly using a dry chemical feeder or dissolved and fed as a solution.

16.2.3.2 Elemental Chlorine

Elemental chlorine is a very hazardous substance that can cause skin irritation, nausea, and death if inhaled in high concentrations. Yellow-green color in gas form and amber-colored in liquid form, chlorine is 2.5 times heavier than air and is non-flammable but is a very strong oxidizing agent. It is available in several different-sized containers (100-, 150-, and 2,000-lb containers or 55- to 90-ton railroad tank cars). The pressurized containers normally contain approximately 80% liquid chlorine and 20% gas. Although chlorine can be fed directly into the wastewater, most facilities dissolve the chlorine gas in water to reduce safety and environmental risks and facilitate movement to the point of application.

Based on water quality standards, total residual limitations on chlorine are

- fresh water—less than 11 ppb total residual chlorine
- estuaries—less that 7.5 ppb for halogen-produced oxidants
- endangered species—use of chlorine is prohibited

16.2.4 CHLORINATION EQUIPMENT

16.2.4.1 Hypochlorite Systems

Depending on the form of hypochlorite selected for use, special equipment that will control the addition of hypochlorite to the wastewater is required. Liquid forms require the use of metering pumps, which can deliver varying flows of hypochlorite solution. Dry chemicals require the use of a feed system designed to provide variable doses of the form used. The tablet form of hypochlorite requires the use of a tablet chlorinator designed specifically to provide the desired dose of chlorine. The hypochlorite solution or dry feed systems dispense the hypochlorite, which is then mixed with the flow. The treated wastewater then enters the contact tank to provide the required contact time.

16.2.4.2 Chlorine Systems

Because of the potential hazards associated with the use of chlorine, the equipment requirements are significantly greater than those associated with hypochlorite use. The system most widely used is a solution feed system. In this system, chlorine is removed from the container at a flow rate controlled by a variable orifice. Water moving through the chlorine injector creates a vacuum that draws the chlorine gas to the injector and mixes it with the water. The chlorine gas reacts with the water to form hypochlorous and hydrochloric acid. The solution is then piped to the chlorine contact tank and dispersed into the wastewater through a diffuser.

Larger facilities may withdraw the liquid form of chlorine and use evaporators (heaters) to convert to the gas form. Small facilities will normally draw the gas form of chlorine from the cylinder. As gas is withdrawn, liquid will be converted to the gas form. This requires heat energy and may result in chlorine line freeze-up if the withdrawal rate exceeds the available energy levels.

16.2.5 CHLORINATION: OPERATION

In either type of system, normal operation requires adjustment of feed rates to ensure the required residual levels are maintained. This normally requires chlorine residual testing and adjustment based upon the results of the test. Other activities include removal of accumulated solids from the contact tank, collection of bacteriological samples to evaluate process performance, and maintenance of safety equipment (respirator-air pack, safety lines, etc.).

Hypochlorite operation may also include make-up solution (solution feed systems), adding powder or pellets to the dry chemical feeder or tablets to the tablet chlorinator.

Chlorine operations include adjusting chlorinator feed rates, inspecting mechanical equipment, testing for leaks using ammonia swab (white smoke means leaks), changing containers (requires more than one person for safety), and adjusting the injector water feed rate when required.

Chlorination requires routine testing of plant effluent for total residual chlorine and may also require collection and analysis of samples to determine the fecal coliform concentration in the effluent.

16.2.6 CHLORINATION ENVIRONMENTAL HAZARDS AND SAFETY

Chlorine is an extremely toxic substance that can, when released to the environment, cause severe damage. For this reason, most state regulatory agencies have established a chlorine water quality standard (e.g., in Virginia, 0.011 mg/L in fresh waters for total residual chlorine and 0.0075 mg/L for chlorine-produced oxidants in saline waters). Studies have indicated that above these levels, chlorine can reduce shellfish growth and destroy sensitive aquatic organisms. This standard has resulted in many treatment facilities being required to add an additional process to remove the chlorine prior to discharge. The process, known as dechlorination (see Section 16.3), uses chemicals that react quickly with chlorine to convert it to a less harmful form.

Elemental chlorine is a chemical with potentially fatal hazards associated with it. For this reason, the transport, storage, and use of chlorine are regulated by many different state and federal agencies. All people required to work with chlorine should be trained in proper handling techniques to ensure that all procedures for storage, transport, handling, and use of chlorine are in compliance with appropriate state and federal regulations.

16.2.6.1 Chlorine: Safe Work Practices[2]

Because of the inherent dangers involved with handling chlorine, each facility using chlorine (for any reason) should ensure that a written safe work practice is in place and is followed by plant operators. A sample safe work practice for handling chlorine is provided in the following.

Work: Chemical Handling—Chlorine

Practice:

(1) Plant personnel *must* be trained and instructed on the use and handling of chlorine, chlorine equipment, chlorine emergency repair kits, and other chlorine emergency procedures.
(2) Use extreme care and caution when handling chlorine.
(3) Lift chlorine cylinders only with an approved and load-tested device.
(4) Secure chlorine cylinders into position immediately. *Never* leave a cylinder suspended.
(5) Avoid dropping chlorine cylinders.
(6) Avoid banging chlorine cylinders into other objects.
(7) Store 1-ton chlorine cylinders in a cool dry place away from direct sunlight or heating units. Railroad tank cars are direct-sunlight compensated.

---

[2]Taken from F. R. Spellman, *Safe Work Practices for Wastewater Treatment Plants.* Lancaster, PA: Technomic Publishing Company, Inc., 177–179, 1996.

(8) Store 1-ton chlorine cylinders on their sides only (horizontally).

(9) Do not stack unused or used chlorine cylinders.

(10) Provide positive ventilation to the chlorine storage area and chlorinator room.

(11) *Always* keep chlorine cylinders at ambient temperature. *Never* apply direct flame to a chlorine cylinder.

(12) Use the oldest chlorine cylinder in stock first.

(13) Always keep valve protection hoods in place until the chlorine cylinders are ready for connection.

(14) Except to repair a leak, do not tamper with the fusible plugs on chlorine cylinders.

(15) Wear a self-contained breathing apparatus (SCBA) whenever changing a chlorine cylinder and have at least one other person with a standby SCBA unit outside the immediate area.

(16) Inspect all threads and surfaces of a chlorine cylinder.

(17) Use new lead gaskets each time a chlorine cylinder connection is made.

(18) Use only the specified wrench to operate chlorine cylinder valves.

(19) Open chlorine cylinder valves slowly, no more than one full turn.

(20) Do not hammer, bang, or force chlorine cylinder valves under any circumstances.

(21) Check for chlorine leaks as soon as the chlorine cylinder connection is made. Leaks are checked for by gently expelling ammonia mist from a plastic squeeze bottle filled with approximately 2 ounces of liquid ammonia solution. Do not put liquid ammonia on valves or equipment.

(22) Correct all minor chlorine leaks at the chlorine cylinder connection immediately.

(23) Except for automatic systems, draw chlorine from only one manifolded chlorine cylinder at a time. *Never* simultaneously open two or more chlorine cylinders connected to a common manifold pulling liquid chlorine. Two or more cylinders connected to a common manifold pulling gaseous chlorine is acceptable.

(24) Wear SCBA and chemical protective clothing covering face, arms, and hands before entering an enclosed chlorine area to investigate a chlorine odor or chlorine leak—two-person rule required.

(25) Provide positive ventilation to a contaminated chlorine atmosphere before entering whenever possible.

(26) Have at least two personnel present before entering a chlorine atmosphere: One person to enter the chlorine atmosphere, the other to observe in the event of an emergency. *Never* enter a chlorine atmosphere unattended. Remember: OSHA mandates that only fully qualified Level III HAZMAT responders are authorized to aggressively attack a hazardous materials leak such as chlorine.

(27) Use supplied-air breathing equipment when entering a chlorine atmosphere. *Never* use canister-type gas masks when entering a chlorine atmosphere.

(28) Ensure that supplied-air breathing apparatus has been properly maintained in accordance with the plant's Self-Contained Breathing Apparatus Inspection Guidelines as specified in the plant's Respiratory Protection Program.

(29) Stay upwind from all chlorine leak danger areas unless involved with making repairs. Look to plant windsocks for wind direction.

(30) Contact trained plant personnel to repair chlorine leaks.

(31) Roll uncontrollable leaking chlorine cylinders so that the chlorine escapes as a gas, not as a liquid.

(32) Stop leaking chlorine cylinders or leaking chlorine equipment (by closing off valve(s) if possible) prior to attempting repair.

(33) Connect uncontrollable leaking chlorine cylinders to the chlorination equipment, and feed the maximum chlorine feed rate possible.

(34) Keep leaking chlorine cylinders at the plant site. Chlorine cylinders received at the plant site must be inspected for leaks prior to taking delivery from the shipper. *Never* ship a leaking chlorine cylinder back to the supplier after it has been accepted (bill of lading has been signed by plant personnel) from the shipper; instead, repair or stop the leak first.

(35) Keep moisture away from a chlorine leak. *Never* put water onto a chlorine leak.

(36) Call the fire department or rescue squad if a person is incapacitated by chlorine.

(37) Administer CPR (use barrier mask if possible) immediately to a person who has been incapacitated by chlorine.

(38) Breathe shallow rather than deep if exposed to chlorine without the appropriate respiratory protection.

(39) Place a person who does not have difficulty breathing and is heavily contaminated with chlorine into a deluge shower. Remove his or her clothing under the water and flush all body parts that were exposed to chlorine.

(40) Flush eyes contaminated with chlorine with copious quantities of lukewarm running water for at least 15 minutes.

(41) Drink milk if throat is irritated by chlorine.

(42) *Never* store other materials in chlorine cylinder storage areas; substances like acetylene and propane are not compatible with chlorine.

16.3 PROCESS CALCULATIONS

There are several calculations that may be useful in operating a chlorination system. Many of these calculations are discussed and illustrated in this section.

16.3.1 CHLORINE DEMAND

Chlorine demand is the amount of chlorine in milligrams per liter that must be added to the wastewater to complete all of the chemical reactions that must occur prior to producing a residual.

$$\text{Chlorine Demand} = \text{Chlorine Dose, mg/L} - \text{Chlorine Residual, mg/L} \qquad (16.1)$$

Example 16.1

Problem:

The plant effluent currently requires a chlorine dose of 7.1 mg/L to produce the required 1.0 mg/L chlorine residual in the chlorine contact tank. What is the chlorine demand in milligrams per liter?

Solution:

$$\text{Chlorine Demand mg/L} = 7.1\,\text{mg/L} - 1.0\,\text{mg/L} = 6.1\,\text{mg/L}$$

16.3.2 CHLORINE FEED RATE

Chlorine feed rate is the amount of chlorine added to the wastewater in pounds per day.

$$\text{Chlorine Feed Rate} = \text{Dose, mg/L} \times \text{Flow, MGD} \times 8.34 \text{ lb/mg/L/MG} \quad (16.2)$$

Example 16.2

Problem:

The current chlorine dose is 5.55 mg/L. What is the feed rate in pounds per day if the flow is 22.89 MGD?

Solution:

$$\text{Feed, lb/day} = 5.55 \text{ mg/L} \times 22.89 \text{ MGD} \times 8.34 \text{ lb/mg/L/MG}$$

$$= 1{,}060 \text{ lb/day}$$

16.3.3 CHLORINE DOSE

Chlorine dose is the concentration of chlorine being added to the wastewater. It is expressed in milligrams per liter.

$$\text{Dose, mg/L} = \frac{\text{Chlorine Feed Rate in Pounds/Day}}{\text{Flow in Million Gallons/Day} \times 8.34 \text{ lb/mg/L/MG}} \quad (16.3)$$

Example 16.3

Problem:

Three hundred twenty (320) pounds of chlorine are added per day to a wastewater flow of 5.60 MGD. What is the chlorine dose in milligrams per liter?

Solution:

$$\text{Dose, mg/L} = \frac{320 \text{ lb/day}}{5.60 \text{ MGD} \times 8.34 \text{ lb/mg/L/MG}} = 6.9 \text{ mg/L}$$

16.3.4 AVAILABLE CHLORINE

When hypochlorite forms of chlorine are used, the available chlorine is listed on the label. In these cases, the amount of chemical added must be converted to the actual amount of chlorine using the following calculation.

$$\text{Available Chlorine} = \text{Amount of Hypochlorite} \times \% \text{ Available Chlorine} \quad (16.4)$$

Example 16.4

Problem:

The calcium hypochlorite used for chlorination contains 62.5% available chlorine. How many pounds of chlorine are added to the plant effluent if the current feed rate is 30 lb of calcium hypochlorite per day?

Solution:

$$\text{Quantity of Chlorine} = 30\,\text{lb} \times 0.625 = 18.75\,\text{lb chlorine}$$

16.3.5 REQUIRED QUANTITY OF DRY HYPOCHLORITE

This calculation is used to determine the amount of hypochlorite needed to achieve the desired dose of chlorine.

$$\text{Hypochlorite Quantity, lb/day} = \frac{\text{Required Chlorine Dose, mg/L} \times \text{Flow, MGD} \times 8.34\,\text{lb/mg/L/MG}}{\%\text{ Available Chlorine}} \quad (16.5)$$

Example 16.5

Problem:

The laboratory reports that the chlorine dose required to maintain the desired residual level is 8.5 mg/L. Today's flow rate is 3.25 MGD. The hypochlorite powder used for disinfection is 70% available chlorine. How many pounds of hypochlorite must be used?

Solution:

$$\text{Hypochlorite Quantity} = \frac{8.5\,\text{mg/L} \times 3.25\,\text{MGD} \times 8.34\,\text{lb/mg/L/MG}}{0.70}$$

$$= 329\,\text{lb/day}$$

16.3.6 REQUIRED QUANTITY OF LIQUID HYPOCHLORITE

$$\text{Hypochlorite Needed, gal/day} = \frac{\text{Required Chlorine Dose, mg/L} \times \text{Flow, MGD} \times 8.34\,\text{lb/mg/L/MG}}{\%\text{ Available Chlorine} \times 8.34 \times \text{Hypochlorite solution spec. gravity}} \quad (16.6)$$

Example 16.6

Problem:

The chlorine dose is 8.8 mg/L, and the flow rate is 3.28 MGD. The hypochlorite solution is 71% available chlorine and has a specific gravity of 1.25. How many pounds of hypochlorite must be used?

Solution:

$$\text{Hypochlorite Quantity} = \frac{8.8\,\text{mg/L} \times 3.28\,\text{MGD} \times 8.34\,\text{lb/mg/L/MG}}{0.71 \times 8.34\,\text{lb/gal} \times 1.25}$$

$$= 32.5\,\text{gal/day}$$

16.3.7 CHLORINE ORDERING

Because disinfection must be continuous, the supply of chlorine must never be allowed to run out. The following calculation provides a simple method for determining when additional supplies must be ordered. The process consists of three steps:

(1) Step 1: Adjust the flow and use variations if projected changes are provided.
(2) Step 2: If an increase in flow and/or required dosage is projected, current flow rate and/or dose must be adjusted to reflect the projected change.
(3) Step 3:

$$\text{Projected Flow} = \text{Current Flow, MGD} \times (1.0 + \% \text{ Change})$$
$$\text{Projected Dose} = \text{Current Dose, mg/L} \times (1.0 + \% \text{ Change})$$
(16.7)

Example 16.7

Problem:

Based on the available information for the past 12 months, the operator projects that the effluent flow rate will increase by 7.5% during the next year. If the average daily flow has been 4.5 MGD, what will be the projected flow for the next 12 months?

Solution:

$$\text{Projected Flow, MGD} = 4.5 \text{ MGD} \times (1.0 + 0.075) = 4.84 \text{ MGD}$$

Determine the amount of chlorine required for a given period.

$$\text{Chlorine Required} = \text{Feed Rate, lb/day} \times \# \text{ of days required}$$

Example 16.8

Problem:

The plant currently uses 90 lb of chlorine per day. The town wishes to order enough chlorine to supply the plant for four months (assume 31 days/month). How many pounds of chlorine should be ordered to provide the needed supply?

Solution:

$$\text{Chlorine Required} = 90 \text{ lb/day} \times 124 \text{ days} = 11,160 \text{ lb}$$

✓ In some instances, projections for flow or dose changes are not available but the plant operator wishes to include an extra amount of chlorine as a safety factor. This safety factor can be stated as a specific quantity or as a percentage of the projected usage.

Safety factor as a specific quantity can be expressed as follows:

$$\text{Total Required Cl}_2 = \text{Chlorine Required, lb} + \text{Safety Factor}$$

✓ Because chlorine is only shipped in full containers, unless asked specifically for the amount of chlorine actually required or used during a specified period, all decimal parts of a cylinder are rounded up to the next highest number of full cylinders.

16.4 TROUBLESHOOTING OPERATIONAL PROBLEMS

On occasion, operational problems with the plant's disinfection process develop. The wastewater operator must not only be able to recognize these problems but also to correct them. In this section, we point out various problems that can occur with the plant's disinfection process, the causes, and the corrective action(s) that should be taken.

16.4.1 SYMPTOM 1

Coliform count fails to meet required standards for disinfection.

(1) *Cause:* Inadequate chlorination equipment capacity
   Corrective action: Replace equipment as necessary to provide treatment based on maximum flow through the pipe.
(2) *Cause:* Inadequate chlorine residual control
   Corrective action: Use chlorine residual analyzer to monitor and control chlorine dosage automatically.
(3) *Cause:* Short circuiting in chlorine contact chamber
   Corrective action: Install baffling in the chlorine contact chamber; install mixing device in chlorine contact chamber.
(4) *Cause:* Solids build up in contact chamber
   Corrective action: Clean contact chamber.
(5) *Cause:* Chlorine residual is too low
   Corrective action: Increase contact time and/or increase chlorine feed rate.

16.4.2 SYMPTOM 2

Low chlorine gas pressure at the chlorinator.

(1) *Cause:* Insufficient number of cylinders connected to the system
   Corrective action: Connect enough cylinders to system so that feed rate does not exceed recommended withdrawal rate for cylinders.
(2) *Cause:* Stoppage or restriction of flow between cylinders and chlorinator
   Corrective action: Disassemble chlorine header system at point where cooling begins, locate stoppage and clean with solvent.

16.4.3 SYMPTOM 3

No chlorine gas pressure at the chlorinator.

(1) *Cause:* Chlorine cylinders empty or not connected to the system
   Corrective action: Connect cylinders or replace empty cylinders.
(2) *Cause:* Plugged or damaged pressure reducing valve
   Corrective action: Repair reducing valve after shutting cylinder valves and decreasing gas in the header system.

16.4.4 SYMPTOM 4

Chlorinator will not feed any chlorine.

(1) *Cause:* Pressure reducing valve in chlorinator is dirty
   Corrective action: Disassemble chlorinator and clean valve stem and seat; precede valve with filter/sediment trap.
(2) *Cause:* Chlorine cylinder is hotter than chlorine control apparatus (chlorinator)
   Corrective action: Reduce temperature in cylinder area; do not connect a new cylinder that has been sitting in the sun.

16.4.5 SYMPTOM 5

Chlorine gas escaping from the chlorine pressure reducing valve (CPRV).

(1) *Cause:* Main diaphragm of CPRV ruptured
   Corrective action: Disassemble valve and diaphragm; inspect chlorine supply system for moisture intrusion.

16.4.6 SYMPTOM 6

Inability to maintain chlorine feed rate without icing of chlorine system.

(1) *Cause:* Insufficient evaporator capacity
   Corrective action: Reduce feed rate to 75% of evaporator capacity. If this eliminates problem, then main diaphragm of CPRV is ruptured.
(2) *Cause:* External CPRV cartridge is clogged
   Corrective action: Flush and clean cartridge.

16.4.7 SYMPTOM 7

Chlorinator system is unable to maintain sufficient water bath temperature to keep external CPRV open.

(1) *Cause:* Heating element malfunction
   Corrective action: Remove and replace heating element.

16.4.8 SYMPTOM 8

Inability to obtain maximum feed rate from chlorinator.

(1) *Cause:* Inadequate chlorine gas pressure
   Corrective action: Increase pressure, replace empty or low cylinders
(2) *Cause:* Water pump injector clogged with deposits
   Corrective action: Clean injector parts using muriatic acid. Rinse parts with fresh water and place back in service.
(3) *Cause:* Leak in vacuum relief valve
   Corrective action: Disassemble vacuum relief valve and replace all springs.

(4) *Cause:* Vacuum leak in joints, gaskets, tubing, etc., in chlorinator system
Corrective action: Repair all vacuum leaks by tightening joints, replacing gaskets, and replacing tubing and/or compression nuts.

16.4.9 SYMPTOM 9

Inability to maintain adequate chlorine feed rate.

(1) *Cause:* Malfunction or deterioration of chlorine water supply pump
Corrective action: Overhaul pump (if turbine pump is used), try closing valve to maintain proper discharge pressure.

16.4.10 SYMPTOM 10

Chlorine residual too high in plant effluent to meet requirements.

(1) *Cause:* Chlorine residual too high
Corrective action: Install dechlorination facilities.

16.4.11 SYMPTOM 11

Wide variation in chlorine residual produced in the effluent.

(1) *Cause:* Chlorine flow proportion meter capacity inadequate to meet plant flow rates
Corrective action: Replace with higher capacity chlorinator meter.
(2) *Cause:* Malfunctioning controls
Corrective action: Call manufacturer technical representative.
(3) *Cause:* Solids settled in chlorine contact chamber
Corrective action: Clean chlorine contact tank.
(4) *Cause:* Flow proportioning control device not zeroed or spanned correctly
Corrective action: Re-zero and span the device in accordance with manufacturer's instructions.

16.4.12 SYMPTOM 12

Unable to obtain chlorine residual.

(1) *Cause:* High chemical demand
Corrective action: Locate and correct the source of the high demand.
(2) *Cause:* Test interference
Corrective action: Add sulfuric acid to samples to reduce interference.

16.4.13 SYMPTOM 13

Chlorine residual analyzer, recorder, controller does not control chlorine residual properly.

(1) *Cause:* Electrodes fouled
Corrective action: Clean electrodes.
(2) *Cause:* Loop time is too long
Corrective action: Reduce control loop time by
- moving the injector closer to the point of application

- increasing the velocity in the sample line to the analyzer
- moving the cell closer to the sample point
- move the sample point closer to the point of application

(3) *Cause:* Insufficient potassium iodide being added for the amount of residual being measured

Corrective action: Adjust potassium iodide feed to correspond with the chlorine residual being measured.

(4) *Cause:* Buffer additive system is malfunctioning

Corrective action: Repair buffer additive system.

(5) *Cause:* Malfunctioning of analyzer cell

Corrective action: Call authorized service personnel to repair electrical components.

(6) *Cause:* Poor mixing of chlorine at point of application

Corrective action: Install mixing device to cause turbulence at point of application.

(7) *Cause:* Rotameter tube range is improperly set

Corrective action: Replace rotameter with a proper range of feed rate.

16.5 DECHLORINATION

The purpose of dechlorination is to remove chlorine and reaction products (chloramines) before the treated wastestream is discharged into its receiving waters. Dechlorination follows chlorination—usually at the end of the contact tank. Sulfur dioxide gas, sodium sulfate, sodium metabisulfate, or sodium bisulfate are the chemicals used to dechlorinate. No matter which chemical is used to dechlorinate, its reaction with chlorine is instantaneous.

16.6 REFERENCE

Spellman, F. R., *Safe Work Practices for Wastewater Treatment Plants*. Lancaster, PA: Technomic Publishing Company, Inc., 1996.

16.7 CHAPTER REVIEW QUESTIONS

16-1 Explain the difference between disinfection and sterilization.

16-2 To be effective enough, chlorine must be added to satisfy the _____ and produce a _____ mg/L _____ for at least _____ minutes at design flow rates.

16-3 Elemental chlorine is _____ in color and is _____ times heavier than air.

16-4 Why must you take safety precautions when working with chlorine?

16-5 You are currently adding 400 lb of chlorine per day to a wastewater flow of 5.55 MGD. What is the chlorine dose in mg/L?

16-6 The chlorine dose is 8.22 mg/L. If the residual is 1.10 mg/L, the chlorine demand is _____.

16-7 Why is dechlorination required to be installed (in many facilities) following chlorination for disinfection?

16-8 The plant adds 350 lb per day of dry hypochlorite powder to the plant effluent. The hypochlorite powder is 42% available chlorine. What is the chlorine feed rate in pounds per day?

16-9 The plant uses liquid HTH, which is 69% available chlorine and has a specific gravity of 1.18. The required feed rate to comply with the plant's discharge permit total residual chlorine limit is 290 lb/day. What is the required flow rate for HTH solution in gallons per day?

16-10 The plant currently uses 45.7 lb of chlorine per day. Assuming the chlorine usage will increase by 10% during the next year, how many 2,000-lb cylinders of chlorine will be needed for the year (365 days)?

16-11 Why are chlorine additions to critical waters, such as natural trout streams, prohibited?

CHAPTER 17

Process Residuals (Sludge) Treatment

17.1 INTRODUCTION

THE wastewater treatment unit processes described to this point remove solids and biochemical oxygen demand (BOD) from the wastestream before the liquid effluent is discharged to its receiving waters. What remains to be disposed of is a mixture of solids and wastes, called *process residuals*—more commonly referred to as *sludge*.

✓ *Note:* As pointed out earlier in Section 2.1, sludge is the commonly accepted name for wastewater solids. However, if wastewater sludge is used for beneficial reuse (e.g., as a soil amendment or fertilizer), it is commonly called biosolids.

It is important to note that the most costly and complex aspect of wastewater treatment can be the collection, processing, and disposal of sludge. This is the case because the quantity of sludge produced may be as high as 2% of the original volume of wastewater, depending somewhat on the treatment process being used. Because sludge can be as much as 97% water content, and because the cost of disposal will be related to the volume of sludge being processed, one of the primary purposes or goals (along with stabilizing it so it is no longer objectionable or environmentally damaging) of sludge treatment is to separate as much of the water from the solids as possible. Sludge treatment methods may be designed to accomplish both of these purposes (see Figure 17.1).

✓ Sludge treatment methods are generally divided into three major categories: thickening, stabilization, and dewatering. Many of these processes include complex sludge treatment methods (i.e., heat treatment, vacuum filtration, incineration, and others) not included in Volume 1 because they are not identified as Class IV/Class III or Grade I/II licensee knowledge and skill requirements. Thus, these complex sludge treatment methods are discussed in detail in Volumes 2 and 3.

✓ Again, the information covering sludge treatment methods covered in this chapter is designed to prepare the licensee with the knowledge he or she needs for Class IV/III or Grade I/II licensure.

Before moving on to a discussion of the fundamentals of sludge treatment methods, it is important to begin by covering sludge pumping calculations. Remember, it is difficult (if not impossible) to treat the sludge unless it is pumped to the specific sludge treatment process.

17.2 SLUDGE PUMPING CALCULATIONS

While on shift, wastewater operators are often called upon to make various process control calculations. An important calculation involves sludge pumping. The sludge pumping calculations

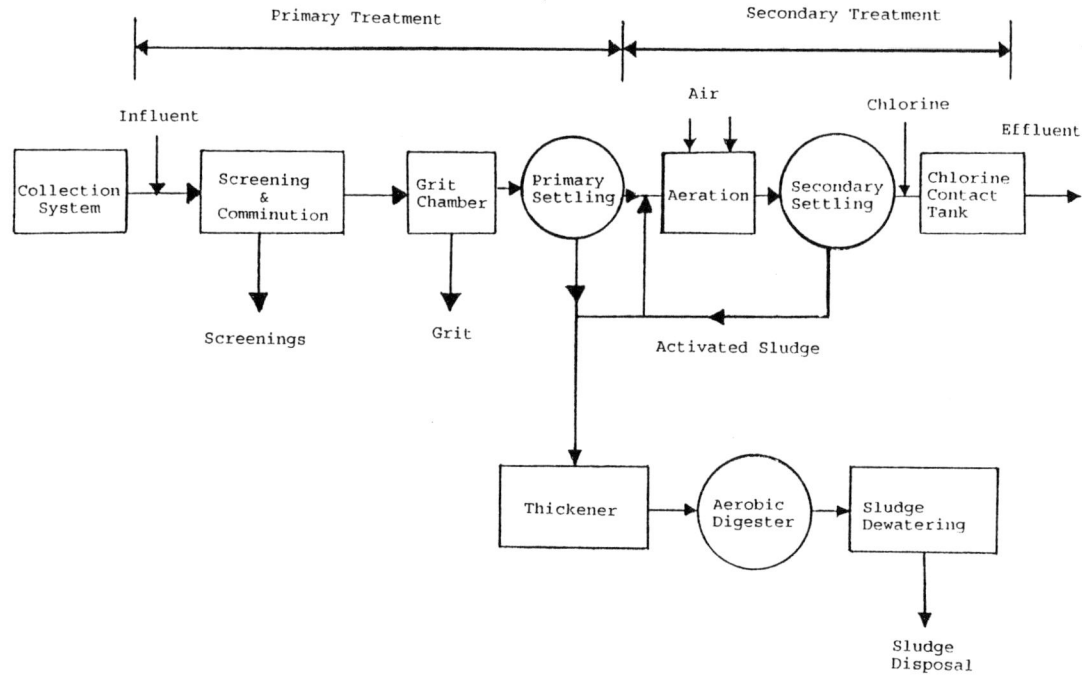

Figure 17.1 Sludge treatment unit processes.

the operator may be required to make during plant operations (and should be known for licensure examinations) are covered in this section.

17.2.1 ESTIMATING DAILY SLUDGE PRODUCTION

The calculation for estimation of the required sludge pumping rate provides a method to establish an initial pumping rate or to evaluate the adequacy of the current withdrawal rate.

$$\text{Estimated Pump Rate} = \frac{(\text{Influent TSS Conc.} - \text{Effluent TSS Conc.}) \times \text{Flow} \times 8.34}{\% \text{ Solids in Sludge} \times 8.34 \times 1{,}440 \text{ min/day}} \quad (17.1)$$

Example 17.1

Problem:

The sludge withdrawn from the primary settling tank contains 1.4% solids. The unit influent contains 285 mg/L TSS, and the effluent contains 140 mg/L TSS. If the influent flow rate is 5.55 MGD, what is the estimated sludge withdrawal rate in gallons per minute (assuming the pump operates continuously)?

Solution:

$$\text{Sludge Rate, gpm} = \frac{(285 \text{ mg/L} - 140 \text{ mg/L}) \times 5.55 \times 8.34}{0.014 \times 8.34 \times 1{,}440 \text{ min/day}} = 40 \text{ gpm}$$

✓ *Note:* The following information is used for examples 17.2–17.6.

Operating Time	15 min/cycle
Frequency	24 times/day
Pump Rate	120 gpm
Solids	3.70%
Volatile Matter	66%

17.2.2 SLUDGE PUMPING TIME

The sludge pumping time is the total time the pump operates during a 24-hour period in minutes.

$$\text{Pump Operating Time} = \text{Time/Cycle, minutes} \times \text{Frequency, cycles/day} \quad (17.2)$$

Example 17.2

Problem:

What is the pump operating time?

Solution:

$$\text{Pump operating time} = 15 \text{ min/hour} \times 24 \text{ (cycles)/day} = 360 \text{ min/day}$$

17.2.3 SLUDGE PUMPED/DAY IN GALLONS

$$\text{Sludge, gpd} = \text{Operating Time, min/day} \times \text{Pump Rate, gpm} \quad (17.3)$$

Example 17.3

Problem:

What is the sludge pumped per day in gallons?

Solution:

$$\text{Sludge, gpd} = 360 \text{ min/day} \times 120 \text{ gpm} = 43{,}200 \text{ gpd}$$

17.2.4 SLUDGE PUMPED/DAY IN POUNDS

$$\text{Sludge, lb/day} = \text{Gallons of Sludge Pumped} \times 8.34 \text{ lb/gal} \quad (17.4)$$

Example 17.4

Problem:

What is the sludge pumped per day in pounds?

Solution:

$$\text{Sludge, lb/day} = 43{,}200 \text{ gal/day} \times 8.34 \text{ lb/gal} = 360{,}300 \text{ lb/day}$$

17.2.5 SOLIDS PUMPED/DAY IN POUNDS

$$\text{Solids Pumped, lb/day} = \text{Sludge Pumped, lb/day} \times \text{\% Solids} \qquad (17.5)$$

Example 17.5

Problem:

What are the solids pumped per day?

Solution:

$$\text{Solids Pumped lb/day} = 360{,}300 \text{ lb/day} \times 0.0370 = 13{,}331 \text{ lb/day}$$

17.2.6 VOLATILE MATTER PUMPED/DAY IN POUNDS

$$\text{Vol. Matter (lb/day)} = \text{Solids Pumped, lb/day} \times \text{\% Volatile Matter} \qquad (17.6)$$

Example 17.6

Problem:

What is the volatile matter in pounds per day?

Solution:

$$\text{Volatile Matter, lb/day} = 13{,}331 \text{ lb/day} \times 0.66 = 8{,}798 \text{ lb/day}$$

✓ *Note:* If you wish to calculate the pounds of solids or the pounds of volatile solids removed per day, the individual equations demonstrated above can be combined into a single calculation.

Solids, lb/day =

Pump Time, min/cycle × Frequency, cycles/day × Rate, gpm × 8.34 lb/gal × solids

$$(17.7)$$

Volatile Matter, lb/day =

Time, min/cyc × Freq. Cyc/day × Rate, gpm × 8.34 × % Solids × % VM

Example 17.7

$$\text{Solids, lb/day} = 15 \text{ min/cyc} \times 24 \text{ cyc/day} \times 120 \text{ gpm} \times 8.34 \times 0.0370$$

$$= 13{,}331 \text{ lb/day}$$

$$\text{VM, lb/day} = 15 \text{ min/cyc} \times 24 \text{ cyc/day} \times 120 \text{ gpm} \times 8.34 \times 0.0370 \times .66$$

$$= 8{,}798 \text{ lb/day}$$

17.2.7 SLUDGE PRODUCTION IN POUNDS/MILLION GALLONS

A common method of expressing sludge production is in pounds of sludge per million gallons of wastewater treated.

$$\text{Sludge, lb/MG} = \frac{\text{Total Sludge Production, lb}}{\text{Total Wastewater Flow, MG}} \quad (17.8)$$

Example 17.8

Problem:

Records show that the plant has produced 85,000 gallons of sludge during the past 30 days. The average daily flow for this period was 1.2 MGD. What was the plant's sludge production in pounds per million gallons?

Solution:

$$\text{Sludge, lb/MG} = \frac{85,000 \text{ gal} \times 8.34 \text{ lb/gal}}{1.2 \text{ MGD} \times 30 \text{ days}} = 19,692 \text{ lb/MG}$$

17.2.8 SLUDGE PRODUCTION IN WET TONS/YEAR

Sludge production can also be expressed in terms of the amount of sludge (water and solids) produced per year. This is normally expressed in wet tons per year.

Sludge, Wet Tons/year =

$$\frac{\text{Sludge Production, lb/MG} \times \text{Average Daily Flow, MGD} \times 365 \text{ days/year}}{2,000 \text{ lb/ton}} \quad (17.9)$$

Example 17.9

Problem:

The plant is currently producing sludge at the rate of 16,500 lb/MG. The current average daily wastewater flow rate is 1.5 MGD. What will be the total amount of sludge produced per year in wet tons per year?

Solution:

$$\text{Sludge, Wet Tons/year} = \frac{16,500 \text{ lb/MG} \times 1.5 \text{ MGD} \times 365 \text{ days/year}}{2,000 \text{ lb/ton}}$$

$$= 4,517 \text{ Wet Tons/year}$$

17.3 SLUDGE THICKENING

Sludge thickening (or concentration) is a unit process used to increase the solids content of the sludge by removing a portion of the liquid fraction. By increasing the solids content, more economical treatment of the sludge can be effected. Sludge thickening processes include the following:

- gravity thickeners
- flotation thickeners
- solids concentrators

17.3.1 GRAVITY THICKENING

Gravity thickening is most effective on primary sludge. In operation, solids are withdrawn from primary treatment (and sometimes secondary treatment) and pumped to the thickener. The solids buildup in the thickener forms a solids blanket on the bottom. The weight of the blanket compresses the solids on the bottom and "squeezes" the water out. By adjusting the blanket thickness, the percent solids in the underflow (solids withdrawn from the bottom of the thickener) can be increased or decreased. The supernatant (clear water) that rises to the surface is returned to the wastewater flow for treatment.

Daily operations of the thickening process include pumping, observation, sampling and testing, process control calculations, maintenance, and housekeeping.

✓ The equipment employed in thickening depends on the specific thickening processes used.

Equipment used for gravity thickening consists of a thickening tank that is similar in design to the settling tank used in primary treatment. Generally, the tank is circular and provides equipment for continuous solids collection. The collector mechanism uses heavier construction than that in a settling tank because the solids being moved are more concentrated. The gravity thickener pumping facilities (i.e., pump and flow measurement) are used for withdrawal of thickened solids.

Performance of gravity thickeners (i.e., the solids concentrations achieved) typically results in producing 8–10% solids from primary underflow, 2–4% solids from waste activated sludge, 7–9% solids from trickling filter residuals, and 4–9% solids from combined primary and secondary residuals.

17.3.2 FLOTATION THICKENING

Flotation thickening is used most efficiently for waste sludges from suspended-growth biological treatment processes, such as the activated sludge process. In operation, recycled water from the flotation thickener is aerated under pressure. During this time, the water absorbs more air than it would under normal pressure. The recycled flow together with chemical additives (if used) is mixed with the flow. When the mixture enters the flotation thickener, the excess air is released in the form of fine bubbles. These bubbles become attached to the solids and lift them toward the surface. The accumulation of solids on the surface is called the float cake. As more solids are added to the bottom of the float cake, it becomes thicker, and water drains from the upper levels of the cake. The solids are then moved up an inclined plane by a scraper and discharged. The supernatant leaves the tank below the surface of the float solids and is recycled or returned to the wastestream for treatment. Typically, flotation thickener performance is 3–5% solids for waste activated sludge with polymer addition and 2–4% solids without polymer addition.

The flotation thickening process requires pressurized air, a vessel for mixing the air with all or part of the process residual flow, a tank for the flotation process to occur, and solids collector mechanisms to remove the float cake (solids) from the top of the tank and accumulated heavy solids from the bottom of the tank. Because the process normally requires chemicals to be added to improve separation, chemical mixing equipment, storage tanks, and metering equipment to dispense the chemicals at the desired dose are required.

17.3.3 SOLIDS CONCENTRATORS

Solids concentrators (gravity belt thickeners) usually consist of a mixing tank, chemical storage and metering equipment, and a moving porous belt. In operation, the process residual flow is chemically treated and then spread evenly over the surface of the moving porous belt. As the flow is carried down the belt (similar to a conveyor belt), the solids are mechanically turned or agitated, and water drains through the belt. This process is primarily used in facilities where space is limited.

17.4 STABILIZATION

The purpose of sludge stabilization is to reduce volume, stabilize the organic matter, and eliminate pathogenic organisms to permit reuse or disposal. The equipment required for stabilization depends on the specific process used. Sludge stabilization processes include the following:

- aerobic digestion
- anaerobic digestion
- composting
- lime stabilization
- wet air oxidation (heat treatment)
- chemical oxidation (chlorine oxidation)
- incineration

17.4.1 AEROBIC DIGESTION

Equipment used for aerobic digestion consists of an aeration tank (digester) that is similar in design to the aeration tank used for the activated sludge process. Either diffused or mechanical aeration equipment is necessary to maintain the aerobic conditions in the tank. Solids and supernatant removal equipment is also required.

In operation, process residuals (sludge) are added to the digester and are aerated to maintain a dissolved oxygen (DO) concentration of 1.0 mg/L. Aeration also ensures that the tank contents are well mixed. Generally, aeration continues for approximately 20 days retention time. Periodically, aeration is stopped, and the solids are allowed to settle. Sludge and the clear liquid supernatant are withdrawn as needed to provide more room in the digester. When no additional volume is available, mixing is stopped for 12–24 hours before solids are withdrawn for disposal. Process control testing should include alkalinity, pH, percent solids, percent volatile solids for influent sludge, supernatant, digested sludge, and digester contents.

Normal operating levels for an aerobic digester are listed in Table 17.1.

A typical operational problem associated with an aerobic digester is pH control. When pH drops, for example, this may indicate normal biological activity or low influent alkalinity. This problem is corrected by adding alkalinity (lime, bicarbonate, etc.).

17.4.1.1 Process Control Calculations: Aerobic Digester

Wastewater operators (who operate aerobic digesters) are required to make certain process control calculations. Moreover, licensing examinations typically include aerobic digester problems for

TABLE 17.1. Normal Operating Levels—Aerobic Digester.

Parameter	Normal Levels
Detention Time, Days	10–20
Volatile Solids Loading lb/ft³/day	0.1–0.3
DO mg/L	1.0
pH	5.9–7.7
Volatile Solids Reduction	40–50%

determining volatile solids loading, digestion time, digester efficiency, and pH adjustment. These process control calculations are explained in the following sections.

17.4.1.1.1 Volatile Solids Loading

Volatile solids loading for the aerobic digester is expressed in pounds of volatile solids entering the digester per day per cubic foot of digester capacity.

$$\text{Volatile Solids Loading} = \frac{\text{Volatile Solids Added, lb/day}}{\text{Digester Volume, ft}^3} \tag{17.10}$$

Example 17.10

Problem:

The aerobic digester is 25 ft in diameter and has an operating depth of 24 ft. The sludge added to the digester daily contains 1,350 lb of volatile solids. What is the volatile solids loading in pounds per day per cubic foot?

Solution:

$$\text{Volatile Solids Loading} = \frac{1{,}350 \text{ lb/day}}{.785 \times 25 \text{ ft} \times 25 \text{ ft} \times 24 \text{ ft}} = 0.11 \text{ lb/day/ft}^3$$

17.4.1.1.2 Digestion Time, Days

Digestion time is the theoretical time the sludge remains in the aerobic digester.

$$\text{Digestion Time, Days} = \frac{\text{Digester Volume, gal}}{\text{Sludge Added, gpd}} \tag{17.11}$$

Example 17.11

Problem:

Digester volume is 240,000 gal. Sludge is being added to the digester at the rate of 13,500 gpd. What is the digestion time in days?

Solution:

$$\text{Digestion Time, Days} = \frac{240{,}000 \text{ gal}}{13{,}500 \text{ gpd}} = 17.8 \text{ days}$$

17.4.1.1.3 Digester Efficiency (Percent Reduction)

To determine digester efficiency or the percent of reduction, a two-step procedure is required. First, the percent volatile matter reduction must be calculated and then the percent moisture reduction.

17.4.1.1.3.1 STEP 1: VOLATILE MATTER

Because of the changes occurring during sludge digestion, the calculation used to determine percent volatile matter reduction is more complicated.

$$\% \text{ Reduction} = \frac{(\% \text{ Volatile Matter}_{in} - \% \text{ Volatile Matter}_{out}) \times 100}{[\% \text{ Vol. Matter}_{in} - (\% \text{ Vol. Matter}_{in} \times \% \text{ Vol. Matter}_{out})]} \quad (17.12)$$

Example 17.12

Problem:

Using the digester data provided below, determine the percent volatile matter reduction for the digester.

| Raw Sludge Volatile Matter | 71% |
| Digested Sludge Volatile Matter | 53% |

Solution:

$$\% \text{ Vol. Matter Reduction} = \frac{(0.71 \times 0.53) \times 100}{[0.71 - (0.71 \times 0.53)]} = 53.9 \text{ or } 54\%$$

17.4.1.1.3.2 STEP 2: MOISTURE REDUCTION

$$\% \text{ Moisture Reduction} = \frac{(\% \text{ Moisture}_{in} - \% \text{ Moisture}_{out}) \times 100}{[\% \text{ Moisture}_{in} - (\% \text{ Moisture}_{in} \times \% \text{ Moisture}_{out})]} \quad (17.13)$$

Example 17.13

Problem:

Using the digester data provided below, determine the percent moisture reduction for the digester.

✓ *Note:* % Moisture = 100% − Percent Solids

Solution:

Raw Sludge	% Solids	6%
	% Moisture	94% (100% − 6%)
Digested Sludge	% Solids	15%
	% Moisture	85% (100% − 15%)

$$\% \text{ Reduction} = \frac{(0.94 - 0.85) \times 100}{[0.94 - (0.94 \times 0.85)]} = 64\%$$

17.4.1.1.4 pH Adjustment

Occasionally, the pH of the aerobic digester will fall below the levels required for good biological activity. When this occurs, the operator must perform a laboratory test to determine the amount of alkalinity required to raise the pH to the desired level. The results of the lab test must then be converted to the actual quantity of chemical (usually lime) required by the digester.

$$\text{Chemical Required, lb} = \frac{\text{Chemical Used in Lab Test, mg}}{\text{Sample Volume, Liters}} \times \text{Dig. Vol. MG} \times 8.34 \qquad (17.14)$$

Example 17.14

Problem:

The lab reports that it took 225 mg of lime to increase pH of a 1-L sample of the aerobic digester contents to pH 7.2. The digester volume is 240,000 gal. How many pounds of lime will be required to increase the digester pH to 7.2?

Solution:

$$\text{Chemical Required, lb} = \frac{225 \text{ mg} \times 240,000 \text{ gal} \times 3.785 \text{ L/gal}}{1 \text{ L} \times 454 \text{ g/lb} \times 1,000 \text{ mg/g}} = 450 \text{ lb}$$

17.4.2 ANAEROBIC DIGESTION

Anaerobic digestion is the traditional method of sludge stabilization. It involves using bacteria that thrive in the absence of oxygen and is slower than aerobic digestion, but has the advantage that only a small percentage of the wastes are converted into new bacterial cells. Instead, most of the organics are converted into carbon dioxide and methane gas.

✓ *Note:* In an anaerobic digester, the entrance of air should be prevented because of the potential for air mixed with the gas produced in the digester, which could create an explosive mixture.

Equipment used in anaerobic digestion includes a sealed digestion tank with either a fixed or a floating cover, heating and mixing equipment, gas storage tanks, solids and supernatant withdrawal equipment, and safety equipment (e.g., vacuum relief, pressure relief, flame traps, explosion proof electrical equipment).

In operation, process residual (thickened or unthickened sludge) is pumped into the sealed digester. The organic matter digests anaerobically by a two-stage process. Sugars, starches, and carbohydrates are converted to volatile acids, carbon dioxide, and hydrogen sulfide. The volatile acids are then converted to methane gas. This operation can occur in a single tank (single stage) or in two tanks (two stage). In a single-stage system, supernatant and/or digested solids must be removed whenever flow is added. In a two-stage operation, solids and liquids from the first stage flow into the second stage each time fresh solids are added. Supernatant is withdrawn from the second stage to provide additional treatment space. Periodically, solids are withdrawn for dewatering or disposal. The methane gas produced in the process may be used for many plant activities.

✓ *Note:* The primary purpose of a secondary digester is to allow for solids separation.

Various performance factors affect the operation of the anaerobic digester. For example, percent volatile matter in raw sludge, digester temperature, mixing, volatile acids/alkalinity ratio, feed rate, percent solids in raw sludge, and pH are all important operational parameters that the operator must monitor.

Along with being able to recognize normal/abnormal anaerobic digester performance parameters, wastewater operators must also know and understand normal operating procedures. Normal operating procedures include sludge additions, supernatant withdrawal, sludge withdrawal, pH control, temperature control, mixing, and safety requirements. Important performance parameters are listed in Table 17.2.

17.4.2.1 Sludge Additions

Sludge must be pumped (in small amounts) several times each day to achieve the desired organic loading and optimum performance.

✓ *Note:* Keep in mind that in fixed cover operations, additions must be balanced by withdrawals. If not, structural damage occurs.

17.4.2.2 Supernatant Withdrawal

Supernatant withdrawal must be controlled for maximum sludge retention time. When sampling, sample all drawoff points and select level with the best quality.

17.4.2.3 Sludge Withdrawal

Digested sludge is withdrawn only when necessary—always leave at least 25% seed.

17.4.2.4 pH Control

pH should be adjusted to maintain 6.8 to 7.2 pH by adjusting feed rate, sludge withdrawal, or alkalinity additions.

✓ *Note:* The buffer capacity of an anaerobic digester is indicated by the volatile acid/alkalinity relationship. Decreases in alkalinity cause a corresponding increase in ratio.

17.4.2.5 Temperature Control

If the digester is heated, the temperature must be controlled to a normal temperature range of 90–95°F. Never adjust the temperature by more than 1°F per day.

TABLE 17.2. Anaerobic Digester—Sludge Parameters.

Raw Sludge Solids	Impact
<4% Solids	Loss of alkalinity
	Decreased sludge retention time
	Increased heating requirements
	Decreased volatile acid:alk. ratio
4–8% Solids	Normal operation
>8% Solids	Poor mixing
	Organic overloading
	Decreased volatile acid:alk. ratio

17.4.2.6 Mixing

If the digester is equipped with mixers, mixing should be accomplished to ensure organisms are exposed to food materials.

17.4.2.7 Safety

Anaerobic digesters are inherently dangerous—several catastrophic failures have been recorded. To prevent such failures, safety equipment such as pressure relief and vacuum relief valves, flame traps, condensate traps, and gas collection safety devices are installed. It is important that these critical safety devices be checked and maintained for proper operation.

✓ *Note:* Because of the inherent danger involved with working inside anaerobic digesters, they are automatically classified as permit-required confined spaces. Therefore, all operations involving internal entry must be made in accordance with OSHA's confined space entry standard.

17.4.3 PROCESS CONTROL TESTING

During operation, anaerobic digesters must be monitored and tested to ensure proper operation. Testing should be accomplished to determine supernatant pH, volatile acids, alkalinity, BOD or chemical oxygen demand (COD), total solids, and temperature. Sludge (in and out) should be routinely tested for percent solids and percent volatile matter. Normal operating parameters are listed in Table 17.3.

17.4.4 ANAEROBIC DIGESTER: TROUBLESHOOTING

As with all other unit processes, the wastewater operator is expected to recognize problematic symptoms with anaerobic digesters and effect the appropriate corrective action(s). Symptoms, causes, and corrective actions are discussed next.

TABLE 17.3. Anaerobic Digester: Normal Operating Ranges.

Parameter	Normal Range
Sludge Retention Time	
Heated	30–60 days
Unheated	180+ days
Volatile Solids Loading	0.04–0.1 lb VM/day/ft^3
Operating Temperature	
Heated	90–95°F
Unheated	Varies with season
Mixing	
Heated—primary	Yes
Unheated—secondary	No
% Methane in Gas	60–72%
% Carbon Dioxide in Gas	28–40%
pH	6.8–7.2
Volatile Acids:Alkalinity Ratio	≤0.1
Volatile Solids Reduction	40–60%
Moisture Reduction	40–60%

17.4.4.1 SYMPTOM 1

Digester gas production is reduced; pH drops below 6.8; and/or volatile acids/alkalinity ratio increases.

Cause:
(1) Digester souring
(2) Organic overloading
(3) Inadequate mixing
(4) Low alkalinity
(5) Hydraulic overloading
(6) Toxicity
(7) Loss of digestion capacity

Corrective actions: Add alkalinity (digested sludge, lime, etc.); improve temperature control; improve mixing; eliminate toxicity; clean digester.

17.4.4.2 SYMPTOM 2

Gray foam oozing from digester.

Cause:
(1) Rapid gasification
(2) Foam-producing organisms present
(3) Foam-producing chemical present

Corrective Actions: Reduce mixing; reduce feed rate; mix slowly by hand; clean all contaminated equipment.

17.4.5 ANAEROBIC DIGESTER: PROCESS CONTROL CALCULATIONS

Process control calculations involved with anaerobic digester operation include determining the required seed volume, volatile acid to alkalinity ratio, sludge retention time, estimated gas production, volatile matter reduction, and percent moisture reduction in digester sludge. Examples on how to make these calculations are provided in the following sections.

17.4.5.1 Required Seed Volume in Gallons

$$\text{Seed Volume (Gallons)} = \text{Digester Volume} \times \% \text{ Seed} \qquad (17.15)$$

Example 17.15

Problem:

The new digester requires a 25% seed to achieve normal operation within the allotted time. If the digester volume is 266,000 gallons, how many gallons of seed material will be required?

Solution:

$$\text{Seed Volume} = 266{,}000 \times 0.25 = 66{,}500 \text{ gal}$$

17.4.5.2 Volatile Acids-to-Alkalinity Ratio

The volatile acids-to-alkalinity ratio can be used to control operation of an anaerobic digester.

$$\text{Ratio} = \frac{\text{Volatile Acids Concentration}}{\text{Alkalinity Concentration}} \tag{17.16}$$

Example 17.16

Problem:

The digester contains 240 mg/L volatile acids and 1,860 mg/L alkalinity. What is the volatile acids/alkalinity ratio?

Solution:

$$\text{Ratio} = \frac{240 \text{ mg/L}}{1,860 \text{ mg/L}} = 0.13$$

✓ Increases in the ratio normally indicate a potential change in the operating condition of the digester as shown in Table 17.4.

17.4.5.3 Sludge Retention Time

Sludge retention time is the length of time the sludge remains in the digester.

$$\text{SRT, Days} = \frac{\text{Digester Volume in Gallons}}{\text{Sludge Volume Added per Day, gpd}} \tag{17.17}$$

Example 17.17

Problem:

Sludge is added to a 525,000-gallon digester at the rate of 12,250 gal per day.

Solution:

$$\text{SRT} = \frac{525,000 \text{ gal}}{12,250 \text{ gpd}} = 42.9 \text{ days}$$

TABLE 17.4.

Operating Condition	VA/Alkalinity Ratio
Optimum	≤0.1
Acceptable range	0.1–0.3
% Carbon dioxide in gas increases	≥0.5
pH decreases	≥0.8

17.4.5.4 Estimated Gas Production in Cubic Feet/Day

The rate of gas production is normally expressed as the volume of gas (ft^3) produced per pound of volatile matter destroyed. The total cubic feet of gas a digester will produce per day can be calculated by

$$\text{Gas Production (ft}^3\text{)} = \text{Vol. Matter In, lb/day} \times \text{Vol. Matter Red.} \times \text{Prod. Rate ft}^3/\text{lb} \quad (17.18)$$

✓ Multiplying the volatile matter added to the digester per day by the % volatile matter reduction (in decimal percent) gives the amount of volatile matter being destroyed by the digestion process per day.

Example 17.18

Problem:

The digester receives 11,450 lb of volatile matter per day. Currently, the volatile matter reduction achieved by the digester is 52%. The rate of gas production is 11.2 ft^3 of gas per pound of volatile matter destroyed.

Solution:

$$\text{Gas Prod.} = 11{,}450 \text{ lb/day} \times 0.52 \times 11.2 \text{ ft}^3/\text{lb} = 66{,}685 \text{ ft}^3/\text{day}$$

17.4.5.5 Volatile Matter Reduction, Percent

Because of the changes occurring during sludge digestion, the calculation used to determine percent volatile matter reduction is more complicated.

$$\% \text{ Reduction} = \frac{(\% \text{ Volatile Matter}_{in} - \% \text{ Volatile Matter}_{out}) \times 100}{[\% \text{ Volatile Matter}_{in} - (\% \text{ Volatile Matter}_{in} \times \% \text{ Volatile Matter}_{out})]} \quad (17.19)$$

Example 17.19

Using the data provided below, determine the percent volatile matter reduction for the digester.

Raw Sludge Volatile Matter   74%
Digested Sludge Volatile Matter   55%

Solution:

$$\% \text{ Volatile Matter Reduction} = \frac{(0.74 - 0.55) \times 100}{[0.74 - (0.74 \times 0.55)]} = 57\%$$

17.4.5.6 Percent Moisture Reduction in Digested Sludge

$$\% \text{ Moisture Reduction} = \frac{(\% \text{ Moisture}_{in} - \% \text{ Moisture}_{out}) \times 100}{[\% \text{ Moisture}_{in} - (\% \text{ Moisture}_{in} \times \% \text{ Moisture}_{out})]} \quad (17.20)$$

Example 17.20

Problem:

Using the digester data provided below, determine the percent moisture reduction and percent volatile matter reduction for the digester.

Solution:

Raw Sludge Percent Solids   6%
Digested Sludge Percent Solids   14%

✓ % Moisture = 100% − Percent Solids

$$\% \text{ Moisture Reduction} = \frac{(0.94 - 0.86) \times 100}{[0.94 - (0.94 \times 0.86)]} = 61\%$$

17.4.6 OTHER SLUDGE STABILIZATION PROCESSES

Along with aerobic and anaerobic digestion, other sludge stabilization processes include composting, lime stabilization, wet air oxidation, and chemical (chlorine) oxidation. These other stabilization processes are briefly described in this section.

✓ *Note:* A more detailed discussion of each of the following sludge stabilization processes is provided in Volumes 2 and 3 (including incineration).

17.4.6.1 Composting

The purpose of composting sludge is to stabilize the organic matter, reduce volume, and eliminate pathogenic organisms. In a composting operation dewatered solids are usually mixed with a bulking agent (i.e., hardwood chips) and stored until biological stabilization occurs. The composting mixture is ventilated during storage to provide sufficient oxygen for oxidation and to prevent odors. After the solids are stabilized, they are separated from the bulking agent. The composted solids are then stored for curing and are applied to farm lands or other beneficial uses. Expected performance of the composting operation for both percent volatile matter reduction and percent moisture reduction ranges from 40 to 60%+.

17.4.6.2 Lime Stabilization

In lime stabilization, process residuals are mixed with lime to achieve a pH of 12.0. This pH is maintained for at least 2 hours. The treated solids can then be dewatered for disposal or directly land applied.

17.4.6.3 Thermal Treatment

Thermal treatment (or wet air oxidation) subjects sludge to high temperature and pressure in a closed reactor vessel. The high temperature and pressure rupture the cell walls of any microorganisms present in the solids and cause chemical oxidation of the organic matter. This process substantially improves dewatering and reduces the volume of material for disposal. It also produces a very high strength waste that must be returned to the wastewater treatment system for further treatment.

17.4.6.4 Chlorine Oxidation

Chlorine oxidation also occurs in a closed vessel. In this process, chlorine (100–1,000 mg/L) is mixed with a recycled solids flow. The recycled flow and process residual flow are mixed in the reactor. The solids and water are separated after leaving the reactor vessel. The water is returned to the wastewater treatment system, and the treated solids are dewatered for disposal. The main advantage of chlorine oxidation is that it can be operated intermittently. The main disadvantage is production of extremely low pH and high chlorine content in the supernatant.

17.4.7 STABILIZATION OPERATION AND PERFORMANCE

Depending on the stabilization process employed, the operational components vary. In general, operations include pumping, observations, sampling and testing, process control calculations, maintenance, and housekeeping. Performance of the stabilization process will also vary with the type of process used. Generally, stabilization processes can produce 40% to 60% reduction of both volatile matter (organic content) and moisture.

17.5 SLUDGE DEWATERING

Digested sludge removed from the digester is still mostly liquid. Sludge dewatering is used to reduce volume by removing the water to permit easy handling and economical reuse or disposal. Dewatering processes include sand drying beds, vacuum filters, centrifuges, filter presses (belt and plate), and incineration.

✓ *Note:* Because most Class IV/III and Grade I/II licensing examinations only touch lightly on this topic, in Volume 1, we only provide a brief description of the major dewatering processes—more detailed coverage is provided in Volume 3 (for Class II/I and Grade III/IV licensee candidates).

17.5.1 SAND DRYING BEDS

Drying beds have been used successfully for years to dewater sludge. Composed of a sand bed (consisting of a gravel base, underdrains, and 8–12 inches of filter grade sand), drying beds include an inlet pipe, splash pad containment walls, and a system to return filtrate (water) for treatment. In some cases, the sand beds are covered to provide drying solids protection from the elements.

In operation, solids are pumped to the sand bed and are allowed to dry by first draining off excess water through the sand and then by evaporation. This is the simplest and cheapest method for dewatering sludge. Moreover, no special training or expertise is required. However, there is a downside; namely, drying beds require a great deal of manpower to clean beds; they can create odor and insect problems; and they can cause sludge buildup during inclement weather.

17.5.2 VACUUM FILTERS

Vacuum filters have also been used for many years to dewater sludge. The vacuum filter includes filter media (belt, cloth, or metal coils), media support (drum), vacuum system, chemical feed equipment, and conveyor belt(s) to transport the dewatered solids.

In operation, chemically treated solids are pumped to a vat or tank in which a rotating drum is submerged. As the drum rotates, a vacuum is applied to the drum. Solids collect on the media and are held there by the vacuum as the drum rotates out of the tank. The vacuum removes additional water from the captured solids. When solids reach the discharge zone, the vacuum is released, and the dewatered solids are discharged onto a conveyor belt for disposal. The media are then washed prior to returning to the start of the cycle.

17.5.2.1 Process Control Calculations

Probably the most frequent calculation vacuum filter operators have to make is for determining filter yield. Example 17.21 illustrates how this calculation is made.

17.5.2.1.1 Filter Yield (lb/h/ft^2): Vacuum Filter

Example 17.21

Problem:

Thickened thermally conditioned sludge is pumped to a vacuum filter at a rate of 50 gpm. The vacuum area of the filter is 12 ft wide with a drum diameter of 9.8 ft. If the sludge concentration is 12%, what is the filter yield in lb/h/ft^2? Assume the sludge weighs 8.34 lb/gal.

Solution:

First, calculate the filter surface area.

$$\text{Area of a cylinder side} = 3.14 \times \text{Diameter} \times \text{Length}$$

$$= 3.14 \times 9.8 \text{ ft} \times 12 \text{ ft} = 369.3 \text{ ft}^2$$

Next, calculate the pounds of solids per hour.

$$\frac{50 \text{ gpm}}{1 \text{ min}} \times \frac{60 \text{ min}}{1 \text{ h}} \times \frac{8.34 \text{ lb}}{1 \text{ gal}} \times \frac{12\%}{100\%} = 3{,}002.4 \text{ lb}/\text{h}$$

Dividing the two:

$$\frac{3{,}002.4 \text{ lb}/\text{h}}{369.3 \text{ ft}^2} = 8.13 \text{ lb}/\text{h}/\text{ft}^2$$

17.5.3 BELT OR PLATE AND FRAME FILTERS (FILTER PRESSES)

Filter presses (belt or plate and frame types) are also used to dewater sludge. The belt filter includes two or more porous belts, rollers, and related handling systems for chemical makeup and feed and supernatant and solids collection and transport.

The plate and frame filter consists of a support frame, filter plates covered with porous material, hydraulic or mechanical mechanism for pressing plates together, related handling systems for chemical makeup and feed, and supernatant and solids collection and transport.

In operation, the belt filter uses a coagulant (polymer) mixed with the influent solids. The chemically treated solids are discharged between two moving belts. First, water drains from the solids by gravity. Then, as the two belts move between a series of rollers, pressure "squeezes" additional water out of the solids. The solids are then discharged onto a conveyor belt for transport to storage/disposal.

In the plate and frame filter, solids are pumped (sandwiched) between plates. Pressure (200 to 250 psi) is applied to the plates, and water is "squeezed" from the solids. At the end of the cycle, the pressure is released, and, as the plates separate, the solids drop out onto a conveyor belt for transport to storage or disposal.

Filter presses have lower operation and maintenance costs than vacuum filters or centrifuges. They typically produce a good quality cake and can be batch operated. However, construction and installation costs are high. Moreover, chemical addition is required, and the presses must be operated by skilled personnel.

17.5.3.1 Process Control Calculations

As part of the operating routine for filter presses, operators are called upon to make certain process control calculations. The process control calculation most commonly used in operating the belt filter press determines the hydraulic loading rate on the unit. The most commonly used process control calculation used in operation of plate and filter presses determines the pounds of solids pressed per hour. Both of these calculations are demonstrated below.

17.5.3.1.1 Hydraulic Loading Rate: Belt Filter Press

Example 17.22

Problem:

A belt filter press receives a daily sludge flow of 0.30 MG. If the belt is 60 in. wide, what is the hydraulic loading rate on the unit in gallons per minute for each foot of belt width (gpm/ft)?

Solution:

$$\frac{0.30 \text{ MG}}{1 \text{ d}} \times \frac{1{,}000{,}000 \text{ gal}}{1 \text{ MG}} \times \frac{1 \text{ d}}{1{,}440 \text{ min}} = \frac{208.3 \text{ gal}}{1 \text{ min}}$$

$$60 \text{ in.} \times \frac{1 \text{ ft}}{12 \text{ in.}} = 5 \text{ ft}$$

$$\frac{208.3 \text{ gpm}}{5 \text{ ft}} = 41.7 \text{ gpm/ft}$$

17.5.3.1.2 Pounds of Solids Pressed Per Hour: Plate and Frame Press

Example 17.23

Problem:

A plate and frame filter press can process 850 gal of sludge during its 120-min operating cycle. If the sludge concentration is 3.7%, and if the plate surface area is 140 ft^2, how many pounds of solids are pressed per hour for each square foot of plate surface area?

Solution:

$$850 \text{ gal} \times \frac{3.7\%}{100\%} \times \frac{8.34 \text{ lb}}{1 \text{ gal}} = 262.3 \text{ lb}$$

$$\frac{262.3 \text{ lb}}{120 \text{ min}} \times \frac{60 \text{ min}}{1 \text{ h}} = 131.2 \text{ lb/h}$$

$$\frac{131.2 \text{ lb/h}}{140 \text{ ft}^2} = 0.94 \text{ lb/h/ft}^2$$

17.5.4 CENTRIFUGES

Centrifuges have been employed in dewatering operations for several years and appear to be gaining in popularity. Depending on the type of centrifuge used, in addition to centrifuge pumping equipment for solids feed and centrate removal, chemical makeup and feed equipment and support systems for removal of dewatered solids are required.

In operation, the centrifuge spins at a very high speed. The centrifugal force it creates "throws" the solids out of the water. Chemically conditioned solids are pumped into the centrifuge. The spinning action "throws" the solids to the outer wall of the centrifuge. The centrate (water) flows inside the unit to a discharge point. The solids held against the outer wall are scraped to a discharge point by an internal scroll moving slightly faster or slower than the centrifuge speed of rotation.

17.5.5 INCINERATORS

Not surprisingly, incinerators produce the maximum solids and moisture reductions. The equipment required depends on whether the unit is a multiple hearth or fluid-bed incinerator. Generally, the system will require a source of heat to reach ignition temperature, a solids feed system, and ash handling equipment. It is important to note that the system must also include all required equipment (e.g., scrubbers) to achieve compliance with air pollution control requirements.

In operation, solids are pumped to the incinerator. The solids are dried then ignited (burned). As they burn, the organic matter is converted to carbon dioxide and water vapor, and the inorganic matter is left behind as ash or "fixed" solids. The ash is then collected for reuse or disposal.

17.6 CHAPTER REVIEW QUESTIONS

17-1 The sludge pump operates 30 minutes every 3 hours. The pump delivers 65 gpm. If the sludge is 5.2% solids and has a volatile matter content of 66%, how many pounds of volatile solids are removed from the settling tank each day?

17-2 Name three commonly used methods to thicken waste activated sludge.

17-3 Is a gravity thickener better at thickening primary or secondary sludge?

17-4 Name three sludge stabilization processes.

17-5 Name three general ways that sludge can be dewatered.

17-6 What two actions take place in a sludge drying bed?

17-7 The aerobic digester has a volume of 52,000 gal. The laboratory test indicates that 41 mg of lime were required to increase the pH of a 1-L sample of digesting sludge from 6.1 to the desired 7.3. How many pounds of lime must be added to the digester to increase the pH of the unit to 7.3?

17-8 The digester has a volume of 72,000 gal. Sludge is added to the digester at the rate of 2,780 gal/day. What is the sludge retention time in days?

17-9 What is the normal operating temperature of a heated anaerobic digester?

17-10 The supernatant contains 335 mg/L volatile acids and 1,840 mg/L of alkalinity. What is the volatile acids alkalinity ratio?

17-11 The digester is 45 ft in diameter and has a depth of 22 ft. Sludge is pumped to the digester at the rate of 4,800 gal/day. What is the sludge retention time?

17-12 The raw sludge pumped to the digester contains 70% volatile matter. The digested sludge removed from the digester contains 47% volatile matter. What is the percent volatile matter reduction?

17-13 Thickened thermally conditioned sludge is pumped to a vacuum filter at a rate of 30 gpm. The vacuum area of the filter is 10 ft wide with a drum diameter of 8.4 ft. If the sludge concentration is 12%, what is the filter yield in lb/h/ft^2? Assume the sludge weight is 8.34 lb/gal.

17-14 Liquids produced during solids treatment must be

17-15 The purpose(s) of sludge treatment is (are):

CHAPTER 18

Wastewater Sampling and Testing

18.1 INTRODUCTION

WHEN an environmental scientist, biologist, or other technical specialist is planning a study that involves sampling, it is important, before initiating the study, to determine the objectives of sampling. One important consideration is to determine whether sampling will be accomplished at a single point or at isolated points. Additionally, frequency of sampling must be determined. That is, will sampling be accomplished at hourly, daily, weekly, monthly, or even longer intervals? Whatever sampling frequency is chosen, the entire process will probably continue over a protracted period.

When a wastewater operator performs sampling, the decision-making requirements mentioned above have already been accomplished: where the samples are to be taken has been determined; the frequency of sampling has been determined; and, as far as whether or not the sampling will continue over a protracted period, it will—for as long as the wastestream continues to flow and the treatment process is continued.

Wastewater operators are required to take samples and test the samples to monitor overall plant and process performance—to determine the effectiveness of treatment. Though there are only a few process control functions to be actually performed and only minimal analysis is required to monitor and report plant daily performance, the importance of sampling and testing in wastewater treatment operations cannot be overstated.

In the following sections, wastewater sampling and testing requirements, information, and the knowledge required for certification as a Class IV/III or Grade I/II wastewater operator are covered.

18.2 WASTEWATER SAMPLING

In wastewater sampling, the first critical step is to obtain good, valid information by collecting a representative sample.

✓ *Note:* A representative sample is one that has the same chemical and/or biological composition as the wastewater it came from.

The second critical step in sampling is to follow a predetermined, plain-English, well-written sampling protocol.

Though it is true that sample type and collection point must always be based on the test requirements and the information sought, it is also true that basic guidelines should be used for all sampling activities. Thus, the third critical step in sampling is to follow sampling rules. Sampling rules that should be followed anytime sampling is undertaken are listed in the following:

- Samples must be collected from a well-mixed location.
- Sampling points must be clearly marked and easy to reach.

- Safety should always be considered when selecting a sampling point.
- Large, non-representative objects must be discarded.
- No deposits, growths, or floating material should be included in the sample.
- All testing must be started as soon as possible after sample collection.
- Samples containing high concentrations of solids or large particles should be homogenized in a blender.
- Sample bottles and sample storage containers should be made of corrosion-resistant material, have leakproof tops, and be capable of withstanding repeated refrigeration and cleaning after use.
- Each sampling location should have a designated storage container used only for samples from that location.
- Appropriate safety procedures should always be followed when collecting samples (rubber gloves, washing after sampling, remaining within guardrails, etc.).

18.2.1 SAMPLING DEVICES AND CONTAINERS

The tools of trade for sampling performed by wastewater operators (and others) always include sampling devices and containers. It is important to ensure that sampling devices are corrosion resistant, easily cleaned, capable of collecting desired samples safely, and in accordance with test requirements. Whenever possible, a sampling device should be assigned to each sampling point. Sampling equipment must be cleaned on a regular basis to avoid contamination.

✓ Some tests require special equipment to ensure the sample is representative. Dissolved oxygen and fecal coliform sampling require special equipment and/or procedures to prevent collection of non-representative samples.

Sample containers may be specified for a particular test. If no container is specified, borosilicate glass or plastic containers may be used. Sample containers should be clean and free of soap or chemical residues.

18.2.2 SAMPLE TYPES

There are two basic types of samples: grab samples and composite samples. The type of sample used depends on the specific test, the reason the sample is being collected, and the requirements in the plant discharge permit.

18.2.2.1 Grab Samples

A grab sample is a discrete sample collected at one time and one location. They are primarily used for any parameter whose concentration can change quickly (i.e., dissolved oxygen, pH, temperature, total chlorine residual), and they are representative only of the conditions at the time of collection. In some instances (small plant, limited staffing), grab samples for permit-related effluent testing are acceptable.

As stated earlier, grab samples must be used to determine pH, total residual chlorine, dissolved oxygen (DO) and also fecal coliform concentrations. However, grab samples may also be used for any test that does not specifically prohibit their use.

✓ *Note:* Before collecting samples for any test procedure, it is best to review the sampling requirements of the test.

18.2.2.2 Composite Samples

A composite sample consists of a series of individual grab samples taken at specified time intervals and in proportion to flow. The individual grab samples are mixed together in proportion to the flow rate at the time the sample was collected to form the composite sample. The composite sample represents the character of the wastewater over a period of time.

18.2.2.2.1 Composite Sampling Procedure

Because knowledge of the procedure used in processing composite samples is important (a basic requirement) to the wastewater operator, the actual procedure used is covered in this section.

Procedure:

- Determine the total amount of sample required for all tests to be performed on the composite sample (because too little sample destroys the accuracy of test results, be sure to allow sufficient sample volume for duplicates and repeats when needed).
- Determine the treatment system's average daily flow.

  ✓ *Note:* Average daily flow can be determined by using several months of data—which will provide a more representative value.

- Calculate a proportioning factor.

  Proportioning Factor (PF) =

$$\frac{\text{Total Sample Volume Required, mm}}{\text{Number of Samples to be Calculated} \times \text{Average Daily Flow, MGD}} \quad (18.1)$$

  ✓ *Note:* Round the proportioning factor to the nearest 50 units (i.e., 50, 100, 150, etc.) to simplify calculation of the sample volume.

- Collect the individual samples in accordance with the schedule (once/hour, once/15 minutes, etc.).
- Determine flow rate at the time the sample was collected.
- Calculate the specific amount to add to the composite container.

$$\text{Required Volume, mL} = \text{Flow}^T \times \text{PF} \quad (18.2)$$

  T = Time sample was collected

- Mix the individual sample thoroughly, measure the required volume, and add to composite storage container.
- Refrigerate the composite sample throughout collection period.

Example 18.1

Problem:

The effluent testing will require 4,200 mL of sample. The average daily flow is 4.25 MGD. Using the flows given below, calculate the amount of sample to be added at each of the times shown:

Time	Flow, MGD
8 A.M.	3.88 MGD
9 A.M.	4.10 MGD
10 A.M.	5.05 MGD
11 A.M.	5.25 MGD
12 Noon	3.80 MGD
1 P.M.	3.65 MGD
2 P.M.	3.20 MGD
3 P.M.	3.45 MGD
4 P.M.	4.10 MGD

Solution:

$$\text{Proportioning Factor (PF)} = \frac{4{,}200 \text{ mL}}{9 \text{ Samples} \times 4.25 \text{ MGD}}$$

$$= 110 \text{ (round down to 100)}$$

$\text{Volume}_{8 \text{ AM}} = 3.88 \times 100 = 388 \text{ (400) mL}$
$\text{Volume}_{9 \text{ AM}} = 4.10 \times 100 = 410 \text{ (410) mL}$
$\text{Volume}_{10 \text{ AM}} = 5.05 \times 100 = 505 \text{ (500) mL}$
$\text{Volume}_{11 \text{ AM}} = 5.25 \times 100 = 525 \text{ (530) mL}$
$\text{Volume}_{12 \text{ N}} = 3.80 \times 100 = 380 \text{ (380) mL}$
$\text{Volume}_{1 \text{ PM}} = 3.65 \times 100 = 365 \text{ (370) mL}$
$\text{Volume}_{2 \text{ PM}} = 3.20 \times 100 = 320 \text{ (320) mL}$
$\text{Volume}_{3 \text{ PM}} = 3.45 \times 100 = 345 \text{ (350) mL}$
$\text{Volume}_{4 \text{ PM}} = 4.10 \times 100 = 410 \text{ (410) mL}$

18.2.3 SAMPLE PRESERVATION METHODS

Because samples can change very rapidly, some tests (e.g., pH, temperature, total residual chlorine, dissolved oxygen) must always be performed immediately (within 15 minutes of collection). Other tests may include methods for preservation of the sample. The preservation method and the maximum allowable holding time are listed in the federal regulations governing wastewater sampling and testing—Federal Regulation (40 CFR 136) *Guidelines Establishing Test Procedures for the Analysis of Pollutants under the Clean Water Act.*

18.3 WASTEWATER TESTING METHODS

It is important to point out that all wastewater sampling and testing must be performed in accordance with the federal regulation. Moreover, references used for sampling and testing must correspond to those listed in the most current federal regulation. For the majority of tests, the following references are cited:

- *Standard Methods for Examination of Water and Wastewater*, American Public Health Association—American Water Works Association—Water Pollution Control Federation, 15th ed., 1980. Bacteriological Testing Only.
- *Standard Methods for Examination of Water and Wastewater*, 18th ed., American Public Health Association, American Water Works Association—Water Environment Federation, Washington, D.C., 1992.

- *Methods for Chemical Analysis of Water and Wastes*, U.S. Environmental Protection Agency, Environmental Monitoring Systems Laboratory—Cincinnati (EMSL-CK), EPA-600/4-79-020. Revised March 1993 and 1979 (where applicable).
- *Annual Book of ASTM Standards, Section 11, Water and Environmental Technology*, American Society of Testing Materials (ASTM), Philadelphia, PA.

✓ *Note:* Only those test methods specifically cited in the federal regulations are approved methods. Other methods contained in the cited references can be used only if the facility receives a variance from the U.S. Environmental Protection Agency.

18.3.1 TEST METHODS

The following sections provide information on sampling and testing for conventional pollutants (pH, TRC, DO, BOD_5, and TSS). Do not attempt to perform any of these tests based upon the information contained in these sections. For detailed information on the various approved methods, consult the appropriate reference listed in the federal regulations or your National Pollutant Discharge Elimination System (NPDES) permit. For additional information on other training materials and hands-on workshops on sampling and testing procedures, contact your State Department of Environmental Quality Regional Office.

✓ *Note:* Historically, licensing examinations have not included requirements to solve complex laboratory tests but have tested the licensee's knowledge of fundamental practices and procedures; therefore, while it is not our intent to have the users of this handbook commit to memory the following procedures, it is our intent to make them familiar with the various laboratory tests that are currently performed by many wastewater operators.

18.3.1.1 pH Testing

The approved method of pH testing requires the use of a pH meter (electrometric method). Grab samples are taken and tested using standard buffer solutions. Selected buffers should have pH values that bracket the expected pH of samples and should be at least 2 pH units apart (i.e., pH 4, 7, 10).

The equipment used in pH testing is listed below:

- beakers, 50–100 mL capacity
- pH meter, readable to 0.1 pH units with or without automatic temperature compensation (ATC)
- pH electrodes, one glass and one calomel or one combination electrode
- thermometer
- magnetic stirrer (optional)
- stirring bars (optional)

Procedure:

✓ The test procedure provided by the manufacturer for the specific instrument being used should always be followed.

In general, the procedures used will include the following steps:

(1) Turn on the meter, and allow it to warm up.
(2) Standardize first with the buffer closest to the expected pH of the sample (normally pH 7.0).

(3) Check meter using a second buffer that will bracket the expected pH of the sample (normally pH 9 or 10).

✓ The current Standard Methods procedure for pH requires use of a three-buffer standardization procedure. This means that step 3 must be repeated using a third buffer (normally pH 4.0).

(4) Gently agitate the buffer or sample during the measurement.
(5) pH is recorded when meter reading is steady.
(6) Always rinse and blot the electrodes with a soft tissue when changing from one solution to another. *Note:* Do not rub the electrodes.
(7) Always use grab samples for pH determinations.
(8) Always rinse electrodes with laboratory grade water and place in tap water *or* pH 7.0 buffer solution.

18.3.1.2 Total Residual Chlorine Sampling and Analysis

Currently, federal regulations cite six approved methods for determination of total residual chlorine (TRC). These are

- DPD-spectrophotometric
- titrimetric—amperometric direct
- titrimetric—iodometric direct
- titrimetric—iodometric back
    starch iodine endpoint—iodine titrant
    starch iodine endpoint—iodate titrant
    amperometric endpoint
- DPD-FAS titration
- chlorine electrode

All of these test procedures are approved methods and, unless prohibited by the plant's NPDES discharge permit, can be used for effluent testing. Based on current most popular method usage in the United States, discussion is limited to

- DPD-spectrophotometric
- DPD-FAS titration
- titrimetric—amperometric direct

✓ Treatment facilities required to meet "non-detectable" total residual chlorine limitations must use one of the test methods specified in the plant's NPDES discharge permit.

For information on any of the other approved methods, refer to the appropriate reference cited in the federal regulations.

18.3.1.2.1 DPD Spectrophotometric

DPD reacts with chlorine to form a red color. The intensity of the color is directly proportional to the amount of chlorine present. This color intensity is measured using a colorimeter or spectrophotometer. This meter reading can be converted to a chlorine concentration using a graph developed by measuring the color intensity produced by solutions with precisely known concentrations of chlorine. In some cases, spectrophotometer or colorimeters are equipped with scales that display chlorine concentrations directly. In these cases, there is no requirement to prepare a standard reference curve.

If the direct reading colorimeter is not used, chemicals that are required to be used include

- potassium dichromate solution 0.100 N
- potassium iodine crystals
- standard ferrous ammonium sulfate solution 0.00282 N
- concentrated phosphoric acid
- sulfuric acid solution (1+5)
- barium diphenylamine sulfonate 0.1%

If an indicator is not used, DPD indicator and phosphate buffer (DPD prepared indicator—buffer + indicator together) are required.

In conducting the test, a direct readout colorimeter designed to meet the test specifications, *or a* spectrophotometer (wavelength of 515 nm and light path of at least 1 cm), *or* a filter photometer with a filter having maximum transmission in the wavelength range of 490 to 530 nm and a light path of at least 1 cm are required. In addition, for direct readout colorimeter procedures, a sample test vial is required. When the direct readout colorimeter procedure is not used, the equipment required includes:

- 250 mL Erlenmeyer flask
- 10 mL measuring pipets
- 15 mL test tubes
- 1 mL pipets (graduated to 0.1 mL)
- Sample cuvettes with 1 cm light path

✓ A cuvette is a small, often tubular laboratory vessel, often made of glass.

Procedure:

✓ For direct readout colorimeters, follow the procedure supplied by the manufacturer.

Standard procedure using spectrophotometer or colorimeter:

(1) Prepare a standard curve for TRC concentrations from 0.05 to 4.0 mg/L—chlorine versus percent transmittance.

   ✓ *Note:* Instructions on how to prepare the TRC concentration curve or a standard curve are normally included in the spectrophotometer manufacturer's operating instructions.

(2) Calibrate colorimeter in accordance with the manufacturer's instructions using a laboratory-grade water blank.
(3) Add one prepared indicator packet (or tablet) of the appropriate size to match sample volume to a clean test tube or cuvette; or
   - Pipet 0.5 mL phosphate buffer solution.
   - Pipet 0.5 mL DPD indicator solution.
   - Add 0.1 g KI (potassium iodide) crystals to a clean tube or cuvette.
(4) Add 10 mL of sample to the cuvette.
(5) Stopper the cuvette, and swirl to mix the contents well.
(6) Let stand for 2 minutes.
(7) Verify the wavelength of the spectrophotometer or colorimeter, and check and set the 0% *T* using the laboratory-grade water blank.

(8) Place the cuvette in instrument, read %*T*, and record reading.
(9) Determine mg/L TRC from standard curve.

✓ *Note:* Calculations are not required in this test because TRC, mg/L is read directly from the meter or from the graph.

18.3.1.2.2 DPD-FAS Titration

The amount of ferrous ammonium sulfate solution required to just remove the red color from a total residual chlorine sample that has been treated with DPD indicator can be used to determine the concentration of chlorine in the sample. This is known as a titrimetric test procedure.

The chemicals used in the test procedure include the following:

- DPD prepared indicator (buffer and indicator together)
- potassium dichromate solution 0.100 N
- potassium iodide crystals
- standard ferrous ammonium sulfate solution 0.00282 N
- concentrated phosphoric acid
- sulfuric acid solution (1+5)
- barium diphenylamine sulfonate 0.1%

✓ *Note:* DPD indicator and/or phosphate buffer are not required if prepared indicator is used.

The equipment required for this test procedure includes the following:

- 250 mL graduated cylinder
- 5 mL measuring pipets
- 500 mL Erlenmeyer flask
- 50 mL buret (graduated to 0.1 mL)
- magnetic stirrer and stir bars

Procedure:

(1) Add the contents of a prepared indicator packet (or tablet) to the Erlenmeyer flask, or
  - pipet 5 mL phosphate buffer solution into an Erlenmeyer flask
  - pipet 5 mL DPD indicator solution into the flask
  - add 1 g KI crystals to the flask
(2) Add 100 mL of sample to the flask.
(3) Swirl the flask to mix contents.
(4) Let the flask stand for 2 minutes.
(5) Titrate with ferrous ammonium sulfate (FAS) until the red color first disappears.
(6) Record the amount of titrant.

The calculation required in this procedure is

$$\text{TRC, mg/L} = \text{mL of FAS used} \tag{18.3}$$

18.3.1.2.3 Titrimetric—Amperometric Direct Titration

In this test procedure, phenylarsine oxide is added to a treated sample to determine when the test reaction has been completed. The volume of phenylarsine oxide (PAO) used can then be used to calculate the TRC.

The chemicals used include the following:

- phenylarsine oxide solution 0.00564 N
- potassium dichromate solution 0.00564 N
- potassium iodide solution 5%
- acetate buffer solution (pH 4.0)
- standard arsenite solution 0.1 N

Equipment used includes the following:

- 250 mL graduated cylinder
- 5 mL measuring pipets
- amperometric titrator

Procedure:

(1) Prepare amperometric titrator according to manufacturer.
(2) Add 200 mL sample.
(3) Place container on titrator stand and turn on mixer.
(4) Add 1 g KI crystals or 1 mL KI solution.
(5) Pipet 1 mL of pH 4 (acetate) buffer into the container.
(6) Titrate with 0.0056 N PAO.

When conducting the test procedure, as the downscale endpoint is neared, slow titrant addition to 0.1 mL increments, and note titrant volume used after each increment. When no needle movement is noted, the endpoint has been reached. Subtract the final increment from the buret reading to determine the final titrant volume.

For this procedure, the only calculation normally required is

$$\text{TRC, mg/L} = \text{mL PAO used} \qquad (18.4)$$

18.3.1.2.4 Iodometric Direct Titration

In this test, phenylarsine oxide (PAO) is added to a treated sample to a starch endpoint (blue to clear). The results of the titration are used to calculate the TRC of the sample.

Chemicals used include the following:

- phenylarsine oxide solution 0.00564 N
- potassium dichromate solution 0.00564 N
- potassium iodide crystals
- acetate buffer solution (pH 4.0)
- standard arsenite solution 0.1 N
- starch indicator

Equipment used includes the following:

- 250 mL graduated cylinder
- 5 mL measuring pipets
- 500 mL Erlenmeyer flask
- 5 mL volumetric pipet
- 10 mL buret (graduated to 0.01 mL)
- 25 mL buret (graduated to 0.1 mL)
- magnetic stirrer and stir bars

Procedure:

(1) Pipet 4 mL (acetate) buffer into an Erlenmeyer flask.
(2) Add 1 g Kl crystals to the flask.
(3) Add 200 mL, 500 mL, or 1,000 mL of sample.
(4) Titrate with 0.00564 N sodium thiosulfate or PAO to pale yellow color.
(5) Add 1–2 mL of starch solution.
(6) Chlorine titration until the blue color disappears.
(7) Repeat steps 1–3 and steps 5–6 with an appropriate volume of laboratory grade water for a negative blank. If no blue appears at step 5, titrate to the first appearance of blue color with 0.0282 N iodine solution, then back titrate with sodium thiosulfate for a positive blank.

The calculations for this (and similar) procedure may be as simple as shown below:

$$\text{TRC, mg/L} = \text{mL PAO used} \tag{18.5}$$

On the other hand, the calculations required to determine TRC using the iodometric direct titration method (and other methods) may be a bit more complicated as demonstrated by the following equations and examples.

18.3.1.2.4.1 IODOMETRIC DIRECT TITRATION METHOD

$$\text{TRC, mg/L} = \frac{(\text{Tit. used for sample, m/L} \pm \text{Tit. used for Blank}) \times \text{Tit. N} \times 35{,}450}{\text{Sample Volume, mL}} \tag{18.6}$$

✓ *Note:* A positive blank (+) is added to the titrant volume, and a negative blank (−) is subtracted from the titrant volume.

Example 18.2

Problem:

Using the information provided, determine TRC, mg/L:

mL of sample	300 mL
Titrant used for sample	2.8 mL
Titrant used for blank	+0.3 mL
Titrant normality	0.00564 N

Solution:

$$\text{TRC, mg/L} = \frac{(2.8 \text{ mL} + 0.3 \text{ mL}) \times 0.00564 \text{ N} \times 35{,}450}{300 \text{ mL}} = 2.1 \text{ mg/L}$$

18.3.1.2.4.2 IODOMETRIC BACK TITRATION USING IODINE TITRANT

$$\text{TRC, mg/L} = \frac{[\text{PAO Added, mL} - (5 \times \text{Titrant Used for Sample, mL})] \times 200}{\text{Sample Volume, mL}} \tag{18.7}$$

Example 18.3

Problem:

Using the information provided, determine TRC, mg/L:

mL of sample	220 mL
PAO added	4.5 mL
Titrant used for sample	0.5 mL

Solution:

$$\text{TRC, mg/L} = \frac{[4.5\text{ mL} - (5 \times 0.5\text{ mL})] \times 200}{220\text{ mL}} = 1.8\text{ mg/L}$$

✓ *Note:* If the Normality (N) of the titrant used for the sample is not exactly 0.0282 N, a corrective factor must be applied to the equation. The correction factor is computed by dividing the actual normality of the titrant by 0.0282 N.

$$CF = \frac{\text{Titrant Normality}}{0.0282\text{ N}} \quad (18.8)$$

When a correction factor is required, the equation is modified as shown below.

$$\text{TRC, mg/L} = \frac{[\text{PAO Added, mL} - (5 \times \text{Tit. Used for Sample} \times \text{Corr. Factor})] \times 200}{\text{Sample Volume, mL}} \quad (18.9)$$

Let's look at an example where the correction factor is used.

Example 18.4

Problem:

Using the information provided below, determine TRC, mg/L:

mL of sample	200 mL
PAO added	4.0 mL
Titrant used for sample	0.5 mL
Correction factor	0.95

Solution:

$$\text{TRC, mg/L} = \frac{[4\text{ mL} - (5 \times 0.5\text{ mL} \times 0.95)] \times 200}{200\text{ mL}} = 1.6\text{ mg/L}$$

18.3.1.2.4.3 IODOMETRIC BACK TITRATION USING IODATE TITRANT

$$\text{TRC, mg/L} = \frac{(\text{Iodate Added to Blank, mL} - \text{Iodate Added to Sample, mL}) \times 200}{\text{Sample Volume, mL}} \quad (18.10)$$

Example 18.5

Problem:

Using the information provided below, determine TRC, mg/L:

mL of sample	200 mL
Iodate used for blank	9.3 mL
Iodate used for sample	7.0 mL

Solution:

$$\text{TRC, mg/L} = \frac{(9.3 \text{ mL} - 7.0 \text{ mL}) \times 200}{200 \text{ mL}} = 2.3 \text{ mg/L}$$

18.3.1.3 Dissolved Oxygen Testing

As the name implies, the dissolved oxygen (DO) test is the testing procedure used to determine the amount of oxygen dissolved in samples of wastewater. The analysis for DO is a key test in water pollution control activities and waste treatment process control. There are various types of tests that can be run to obtain the amount of DO. In this volume of the handbook, the two approved methods we are concerned with are the dissolved oxygen meter method and azide modification of the Winkler method.

18.3.1.3.1 Dissolved Oxygen Meter Method

If samples are to be collected for analysis in the laboratory, a special APHA sampler, or the equivalent, must be used. This is the case because, if the sample is exposed or mixed with air during collection, test results can change dramatically. Therefore, the sampling device must allow collection of a sample that is not mixed with atmospheric air and allows for at least a 3× bottle overflow (see Figure 18.1).

Again, because the DO level in a sample can change quickly, only grab samples should be used for dissolved oxygen testing. Samples must be tested immediately (within 15 minutes) after collection.

✓ *Note:* Samples collected for analysis using the modified Winkler titration method may be preserved for up to 8 hours by adding 0.7 mL of concentrated sulfuric acid or by adding all the chemicals required by the procedure. Samples collected from the aeration tank of the activated sludge process must be preserved using a solution of copper sulfate-sulfamic acid to inhibit biological activity.

The advantage of using the DO oxygen meter method is that the meter can be used to determine DO concentration directly (see Figure 18.2). In the field, a direct reading can be obtained using a field probe (see Figure 18.3) or by collection of samples for testing in the laboratory using a laboratory probe (see Figure 18.4).

✓ The field probe can be used for laboratory work by placing a stirrer in the bottom of the sample bottle, but the laboratory probe should never be used in any situation where the entire probe might be submerged.

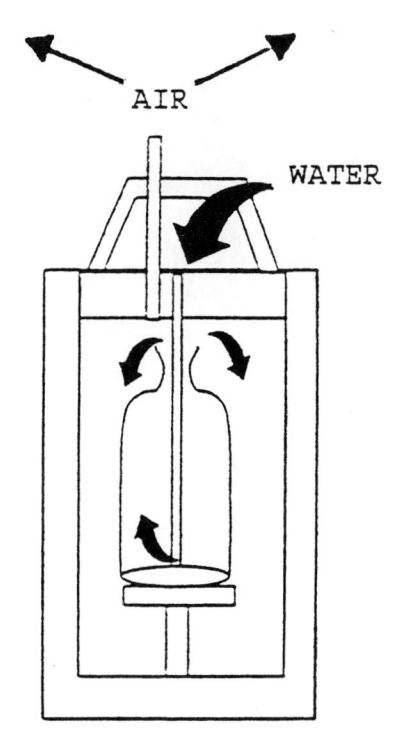

Figure 18.1 To prevent samples being collected without being mixed with air, the special device shown here is used. This device collects the sample below the surface and permits at least three overflows of the sample bottle.

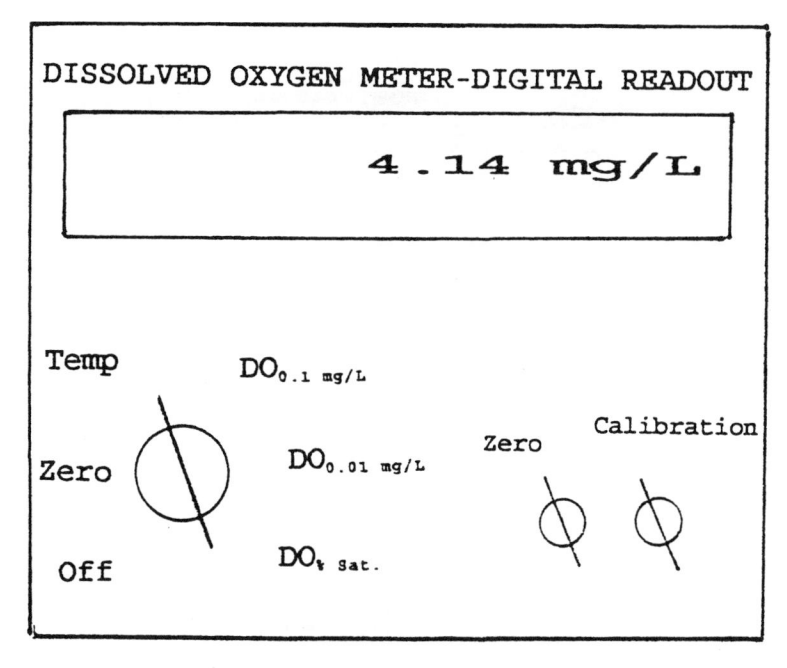

Figure 18.2 Dissolved oxygen meter.

217

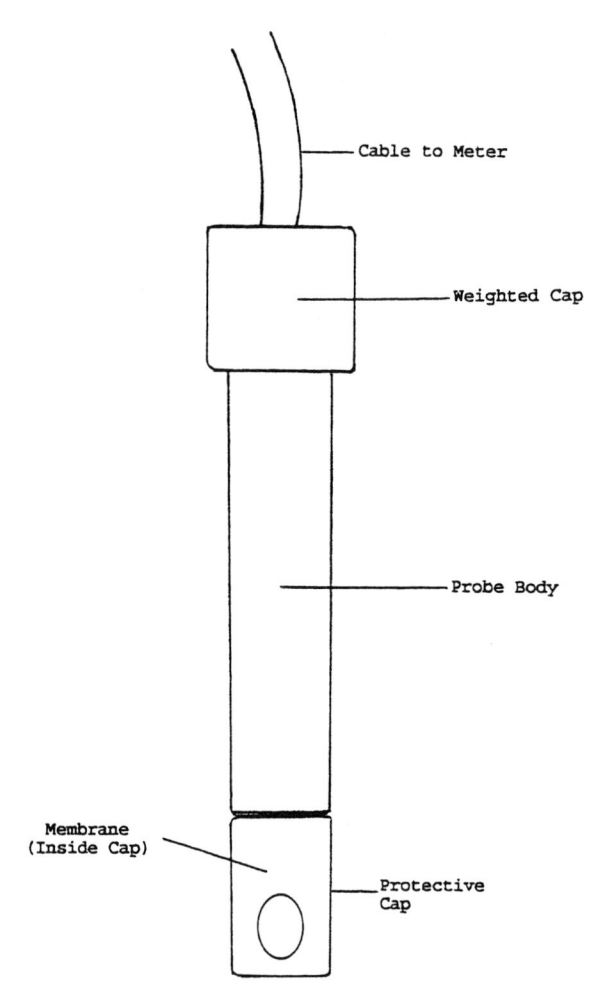

Figure 18.3 Dissolved oxygen-field probe.

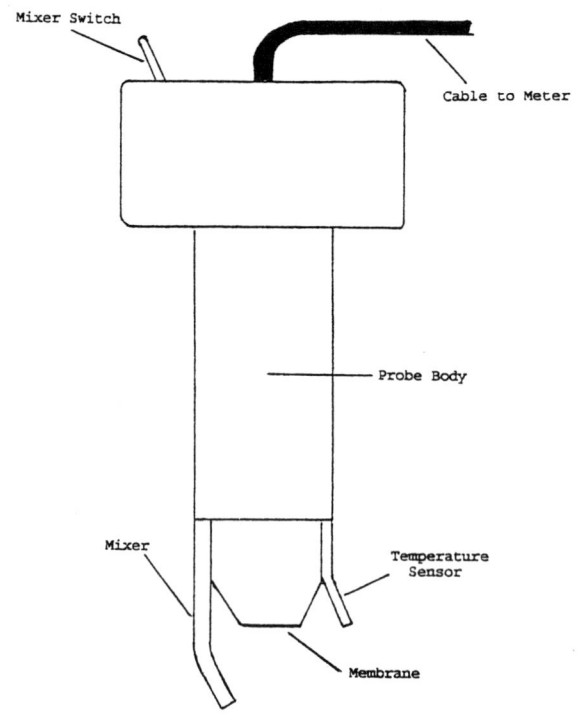

Figure 18.4 Dissolved oxygen-lab probe.

The probe used in the determination of DO consists of two electrodes, a membrane and a membrane filling solution. Oxygen passes through the membrane into the filling solution and causes a change in the electrical current passing between the two electrodes. The change is measured and displayed as the concentration of DO. In order to be accurate, the probe membrane must be in proper operating condition, and the meter must be calibrated prior to use.

The only chemical used in the DO meter method during normal operation is the electrode filling solution. However, in the Winkler DO method, chemicals are required for meter calibration.

Calibration prior to use is important. Both the meter and the probe must be calculated to ensure accurate results. The frequency of calibration is dependent on the frequency of use. For example, if the meter is used once a day, then calibration should be performed prior to use. There are three methods available for calibration: saturated water, saturated air, and the Winkler method. It is important to point out that if the Winkler method is not used for routine calibration method, periodic checks using this method are recommended.

Procedure:

It is important to keep in mind that the meter and probe supplier's operating procedures should always be followed. Normally, the manufacturer's recommended procedure will include the following generalized steps:

(1) Turn DO meter on, and allow 15 minutes for it to warm up.
(2) Turn meter switch to zero, and adjust as needed.
(3) Calibrate meter using the saturated air, saturated water, or Winkler azide procedure for calibration.
(4) Collect sample in 300 mL bottle, or place field electrode directly in stream.
(5) Place laboratory electrode in BOD bottle without trapping air against membrane, and turn on stirrer.
(6) Turn meter switch to temperature setting, and measure temperature.
(7) Turn meter switch to DO mode, and allow 10 seconds for meter reading to stabilize.
(8) Read DO mg/L from meter, and record the results.

No calculation is necessary using this method because results are read directly from the meter.

18.3.1.3.2 Modified Winkler Method (Azide Modification)

In addition to the DO meter method, the azide modification of the Winkler Method can be used to test for dissolved oxygen content in wastewater samples. The azide modification method is best suited for relatively clean waters; otherwise, substances such as color, organics, suspended solids, sulfide, chlorine, and ferrous and ferric iron can interfere with test results. If fresh azide is used, nitrite will not interfere with the test.

In operation, iodine is released in proportion to the amount of DO present in the sample. By using sodium thiosulfate with starch as the indicator, the sample can be titrated to determine the amount of DO present.

Chemicals used include

- manganese sulfate solution
- alkaline azide-iodide solution
- sulfuric acid—concentrated
- starch indicator
- sodium thiosulfate solution 0.025 N, or phenylarsine oxide solution 0.025 N, or potassium biniodate solution 0.025 N
- distilled or deionized water

Equipment used includes

- buret, graduated to 0.1 mL
- buret stand
- 300 mL BOD bottles
- 500 mL Erlenmeyer flasks
- 1.0 mL pipets with elongated tips
- pipet bulb
- 250 mL graduated cylinder
- laboratory-grade water rinse bottle
- magnetic stirrer and stir bars (optional)

Procedure:

(1) Collect sample in a 300 mL BOD bottle.
(2) Add 1 mL manganous sulfate solution at the surface of the liquid.
(3) Add 1 mL alkaline-iodide-azide solution at the surface of the liquid.
(4) Stopper bottle, and mix by inverting the bottle.
(5) Allow the floc to settle halfway in the bottle, remix, and allow to settle again.
(6) Add 1 mL concentrated sulfuric acid at the surface of the liquid.
(7) Restopper bottle, rinse top with laboratory-grade water, and mix until precipitate is dissolved.
(8) The liquid in the bottle should appear clear and have an amber color.
(9) Measure 201 mL from the BOD bottle into an Erlenmeyer flask.
(10) Titrate with 0.025 N PAO or thiosulfate to a pale yellow color, and note the amount of titrant.
(11) Add 1 mL of starch indicator solution.
(12) Titrate until blue color first disappears.
(13) Record total amount of titrant.

18.3.1.3.2.1 CALCULATION

To calculate the DO concentration when the modified Winkler titration method is used:

$$DO, mg/L = \frac{(Buret_{Final}, mL - Buret_{Start}, mL) \times N \times 8,000}{Sample\ Volume, mL} \quad (18.11)$$

✓ Using a 200-mL sample and a 0.025 N (N = Normality of the solution used to titrate the sample) titrant reduces this calculation to

$$DO, mg/L = mL\ Titrant\ Used$$

Example 18.6

Problem:

The operator titrates a 200-mL DO sample. The buret reading at the start of the titration was 0.0 mL. At the end of the titration, the buret read 7.1 mL. The concentration of the titrating solution was 0.025 N. What is the DO concentration in mg/L?

$$DO, mg/L = \frac{(7.1\ mL - 0.0\ mL) \times 0.025 \times 8,000}{200\ mL} = 7.1\ mL$$

18.3.1.4 Biochemical Oxygen Demand Sampling and Analysis

The approved biochemical oxygen demand sampling and analysis procedure measures the DO depletion (biological oxidation of organic matter in the sample) over a 5-day period under controlled conditions (20°C in the dark). The test is performed using a specified incubation time and temperature. Test results are used to determine plant loadings, plant efficiency, and compliance with NPDES effluent limitations. The duration of the test (5 days) makes it difficult to use the data effectively for process control.

The standard BOD test does not differentiate between oxygen used to oxidize organic matter and that used to oxidize organic and ammonia nitrogen to more stable forms. Because many biological treatment plants now control treatment processes to achieve oxidation of the nitrogen compounds, there is a possibility that BOD test results for plant effluent and some process samples may produce BOD rest results based on both carbon and nitrogen oxidation. To avoid this situation, a nitrification inhibitor can be added. When this is done, the test results are known as *carbonaceous BOD* (CBOD). A second uninhibited BOD should also be run whenever CBOD is determined.

When taking a BOD sample, no special sampling container is required. Either a grab or composite sample can be used. BOD_5 samples can be preserved by refrigeration at or below 4°C (not frozen)—composite samples must be refrigerated during collection. Maximum holding time for preserved samples is 48 hours.

Using the incubation or dissolved approved test method, a sample is mixed with dilution water in several different concentrations (dilutions). The dilution water contains nutrients and materials to provide optimum environment. Chemicals used: dissolved oxygen, ferric chloride, magnesium sulfate, calcium chloride, phosphate buffer, and ammonium chloride.

✓ *Note:* Remember all chemicals can be dangerous if not used properly and in accordance with the recommended procedures. Review appropriate sections of the individual chemical materials safety data sheet (MSDS) to determine proper methods for handling and for safety precautions that should be taken.

Sometimes, it is necessary to add (seed) healthy organisms to the sample. The DO of the dilutions and the dilution water is determined. If seed material is used, a series of dilutions of seed material must also be prepared. The dilutions and dilution blanks are incubated in the dark for 5 days at 20°C ± 1°C. At the end of 5 days, the DO of each dilution and the dilution blanks are determined.

For the test results to be valid, certain criteria must be achieved. These test criteria are listed as follows:

- Dilution water blank DO change must be ≤0.2 mg/L.
- Initial DO must be >7.0 mg/L but ≤9.0 mg/L (or saturation at 20°C and test elevation).
- Sample dilution DO depletion must be ≥2.0 mg/L.
- Sample dilution residual DO must be ≥1.0 mg/L.
- Sample dilution initial DO must be ≥7.0 mg/L.
- Seed correction should be ≥0.6 but ≤1.0 mg/L.

The BOD_5 test procedure consists of 10 steps (for unchlorinated water) as shown in Table 18.1.

BOD_5 is calculated individually for all sample dilutions that meet the criteria. Reported result is the average of the BOD_5 of each valid sample dilution.

18.3.1.4.1 BOD₅ Calculation (Unseeded)

Unlike the direct reading instrument used in the DO analysis, BOD results require calculation. There are several criteria used in selecting which BOD_5 dilutions should be used for calculating test results. Consult a laboratory testing reference manual (such as *Standard Methods*) for this information.

TABLE 18.1. BOD₅ Test Procedure.

(1) Fill two bottles with BOD dilution water; insert stoppers.
(2) Place sample in two BOD bottles; fill with dilution water; insert stoppers.
(3) Test for dissolved oxygen (DO).
(4) Incubate for 5 days.
(5) Test for DO.
(6) Add 1 mL MnSO₄ below surface.
(7) Add 1 mL alkaline KI below surface.
(8) Add 1 mL H₂SO₄.
(9) Transfer 203 mL to flask.
(10) Titrate with PAO or thiosulfate.

At the present time, there are two basic calculations for BOD₅. The first is used for samples that have not been seeded. The second must be used whenever BOD₅ samples are seeded. In this section, we illustrate the calculation procedure for unseeded samples.

$$\text{BOD}_5 \text{ (Unseeded)} = \frac{(\text{DO}_{start}, \text{mg/L} - \text{DO}_{final}, \text{mg/L}) \times 300 \text{ mL}}{\text{Sample Volume, mL}} \quad (18.12)$$

Example 18.7

Problem:

The BOD₅ test is completed. Bottle 1 of the test had a DO of 7.1 mg/L at the start of the test. After 5 days, bottle 1 had a DO of 2.9 mg/L. Bottle 1 contained 120 mL of sample.

Solution:

$$\text{BOD}_5 \text{ (Unseeded)} = \frac{(7.1 \text{ mg/L} - 2.9 \text{ mg/L}) \times 300 \text{ mL}}{120 \text{ mL}} = 10.5 \text{ mg/L}$$

18.3.1.4.2 BOD₅ (Seeded)

If the BOD₅ sample has been exposed to conditions that could reduce the number of healthy, active organisms, the sample must be seeded with organisms. Seeding requires use of a correction factor to remove the BOD₅ contribution of the seed material.

$$\text{Seed Correction} = \frac{\text{Seed Material BOD}_5 \times \text{Seed in Dilution, mL}}{300 \text{ mL}} \quad (18.13)$$

$$\text{BOD}_5 \text{ (Seeded)} = \frac{[(\text{DO}_{start}, \text{mg/L} - \text{DO}_{final}, \text{mg/L}) - \text{Seed Corr.}] \times 300}{\text{Sample Volume, mL}} \quad (18.14)$$

Example 18.8

Problem:

Using the data provided below, determine the BOD₅:

BOD$_5$ of Seed Material		90 mg/L
Dilution #1	mL of seed material	3 mL
	mL of sample	100 mL
	Start DO	7.6 mg/L
	Final DO	2.7 mg/L

Solution:

$$\text{Seed Correction} = \frac{90 \text{ mg/L} \times 3 \text{ mL}}{300 \text{ mL}} = 0.90 \text{ mg/L}$$

$$\text{BOD}_5 \text{ (Seeded)} = \frac{[(7.6 \text{ mg/L} - 2.7 \text{ mg/L}) - 0.90] \times 300}{100 \text{ mL}} = 12 \text{ mg/L}$$

18.3.1.5 Solids Sampling and Testing

✓ The solids sampling and testing procedures we are concerned with in Volume 1 are for total suspended solids and volatile suspended solids.

18.3.1.5.1 Total Suspended Solids

The term *solid* means any material suspended or dissolved in wastewater. Although normal domestic wastewater contains a very small amount of solids (usually less than 0.1%), most treatment processes are designed specifically to remove or convert solids to a form that can be removed or discharged without causing environmental harm.

When conducting solids testing, there are many things that affect the accuracy of the test or result in wide variations in results for a single sample. These include the following:

- drying temperature
- length of drying time
- condition of desiccator and desiccant
- non-representative samples' lack of consistency in test procedure
- failure to achieve constant weight prior to calculating results

There are several precautions that can help to increase the reliability of test results:

(1) Use extreme care when measuring samples, weighing materials, and drying or cooling samples.
(2) Check and regulate oven and furnace temperatures frequently to maintain the desired range.
(3) Use an indicator drying agent in the desiccator that changes color when it is no longer good—change or regenerate the desiccant when necessary.
(4) Keep desiccator cover greased with the appropriate type of grease—this will seal the desiccator and prevent moisture from entering the desiccator as the test glassware cools.
(5) Check ceramic glassware for cracks and glass fiber filter for possible holes. A hole in a glass filter will cause solids to pass through and give inaccurate results.
(6) Follow manufacturer's recommendation for care and operation of analytical balances.

In sampling for total suspended solids (TSS), samples may be either grab or composite and can be collected in either glass or plastic containers. TSS samples can be preserved by refrigeration at or below 4°C (not frozen). However, composite samples must be refrigerated during collection. The maximum holding time for preserved samples is 7 days.

Test Procedure:

In short, to conduct a TSS test procedure, a well-mixed measured sample is poured into a filtration apparatus and, with the aid of a vacuum pump or aspirator, is drawn through a preweighed glass fiber filter. After filtration, the glass filter is dried at 103–105°C, cooled and reweighed. The increase in weight of the filter and solids compared to the filter alone represents the total suspended solids.

An example of the specific test procedure used for total suspended solids is given below.

(1) Select a sample volume that will yield between 10 and 200 mg of residue with a filtration time of 10 minutes or less.

- ✓ If filtration time exceeds 10 minutes, increase filter area or decrease volume to reduce filtration time.

- ✓ For non-homogenous samples or samples with very high solids concentrations (i.e., raw wastewater or mixed liquor), use a larger filter to ensure a representative sample volume can be filtered.

(2) Place preweighed glass fiber filter on filtration assembly in a filter flask.
(3) Mix sample well, and measure the selected volume of sample.
(4) Apply suction to filter flask, and wet filter with a small amount of laboratory-grade water to seal it.
(5) Pour the selected sample volume into filtration apparatus.
(6) Draw sample through filter.
(7) Rinse measuring device into filtration apparatus with three successive 10 mL portions of laboratory-grade water. Allow complete drainage between rinsings.
(8) Continue suction for 3 minutes after filtration of final rinse is completed.
(9) Remove the glass filter from the filtration assembly (membrane filter funnel or clean Gooch crucible). If using the large disks and membrane filter assembly, transfer the glass filter to a support (aluminum pan or evaporating dish) for drying.
(10) Place the glass filter with solids and support (pan, dish, or crucible) in a drying oven.
(11) Dry filter and solids to constant weight at 103–105°C (at least 1 hour).
(12) Cool to room temperature in a desiccator.
(13) Weigh the filter, and support and record constant weight in test record.

18.3.1.5.1.1 TSS CALCULATIONS

To determine the total suspended solids concentration in mg/L, we use the following equations:

- to determine weight of dry solids in grams

$$\text{Dry Solids, g} = \text{Wt. of Dry Solids and Filter, g} - \text{Wt. of Dry Filter, g} \quad (18.15)$$

- to determine weight of dry solids in milligrams (mg)

$$\text{Dry Solids, mg} = \text{Wt. of Solids and Filter, g} - \text{Wt. of Dry Filter, g} \quad (18.16)$$

- to determine the TSS concentration in mg/L

$$\text{TSS, mg/L} = \frac{\text{Dry Solids, mg} \times 1{,}000 \text{ mL/L}}{\text{mL sample}} \quad (18.17)$$

Example 18.9

Problem:

Using the data provided below, calculate total suspended solids (TSS):

Sample Volume, mL	250 mL
Weight of Dry Solids and Filter, g	2.305 g
Weight of Dry Filter, g	2.297 g

Solution:

$$\text{Dry Solids, g} = 2.305\,g - 2.297\,g = 0.008\,g$$

$$\text{Dry Solids, mg} = 0.008\,g \times 1{,}000\,mg/g = 8\,mg$$

$$\text{TSS, mg/L} = \frac{8.0 \times 1{,}000\,mL/L}{250\,mL} = 32.0\,mg/L$$

18.3.1.5.2 Volatile Suspended Solids

When the total suspended solids are ignited at $550 \pm 50°C$, the volatile (organic) suspended solids of the sample are converted to water vapor and carbon dioxide and are released to the atmosphere. The solids that remain after the ignition (ash) are the inorganic or fixed solids.

In addition to the equipment and supplies required for the total suspended solids test, you need the following:

- muffle furnace ($550 \pm 50°C$)
- ceramic dishes
- furnace tongs
- insulated gloves

Test Procedure:

(1) Place the weighed filter with solids and support from the total suspended solids test in the muffle furnace.
(2) Ignite filter, solids, and support at $550 \pm 50°C$ for 15–20 minutes.
(3) Remove the ignited solids, filter, and support from the furnace, and partially air cool.
(4) Cool to room temperature in a desiccator.
(5) Weigh ignited solids, filter, and support on an analytical balance.
(6) Record weight of ignited solids, filter, and support.

18.3.1.5.2.1 TOTAL VOLATILE SUSPENDED SOLIDS CALCULATIONS

To calculate total volatile suspended solids (TVSS) requires the following information:

- weight of dry solids, filter, and support in grams
- weight of ignited solids, filter and support in grams

$$\text{Total Volatile Suspended Solids, mg/L} = \frac{(A-C) \times 1{,}000\,mg/g \times 1{,}000\,mL/L}{\text{Sample Volume, mL}} \quad (18.18)$$

A = Weight of Dried Solids, Filter, and Support

C = Weight of Ignited Solids, Filter, and Support

Example 18.10

$$\text{Tot. Vol. Sus. Sol.} = \frac{(1.6530 \text{ g} - 1.6330 \text{ g}) \times 1{,}000 \text{ mg/g} \times 1{,}000 \text{ mL}}{100 \text{ mL}}$$

$$= \frac{0.02 \times 1{,}000{,}000 \text{ mg/L}}{100}$$

$$= 200 \text{ mg/L}$$

✓ Total fixed suspended solids (TFSS) is the difference between the total volatile suspended solids (TVSS) and the total suspended solids (TSS) concentrations.

$$\text{Fixed Sus. Solids} = \text{Total Sus. Solids} - \text{Volatile Sus. Solids} \qquad (18.19)$$

Example: 18.11

Given:

$$\text{Total Suspended Solids} = 202 \text{ mg/L}$$

$$\text{Total Volatile Suspended Solids} = 200 \text{ mg/L}$$

$$\text{Total Fixed Suspended Solids, mg/L} = 202 \text{ mg/L} - 200 \text{ mg/L} = 2 \text{ mg/L}$$

18.4 CHAPTER REVIEW QUESTIONS

18-1 How soon after the sample is collected must the pH be tested?

18-2 The operator titrates a 200 mL dissolved oxygen sample. The buret reading at the start of the titration was 0.0 mL. At the end of the titration, the buret read 7.1 mL. The concentration of the titrating solution was 0.025. What is the DO concentration in mg/L?

18-3 What is a grab sample?

18-4 Dissolved oxygen samples collected from the aeration tank and carried back to the lab for testing must be preserved by adding the following:

18-5 When is it necessary to use a grab sample?

18-6 If a grab sample is to be used to evaluate plant performance, when should the influent and effluent samples be collected?

18-7 What is a composite sample?

18-8 Why is a composite sample more representative of the average characteristics of the wastewater?

18-9 List three rules for sample collection.

18-10 The approved method for pH testing requires the following:

18-11 Who specifies the sample type, preservation method, and test method for effluent samples?

18-12 The *Guidelines Establishing Test Procedures for the Analysis of Pollutants under the Clean Water Act* is a/an _____ regulation.

18-13 The average daily flow is 7.66 MGD, and the effluent testing will require 3,000 mL of sample. What is the proportioning factor, if a 24-hour composite sample is to be collected?

18-14 The proportioning factor is 100, and the flow at the time the 7 A.M. sample is collected is 4.66 MGD. How many milliliters of sample should be added to the composite sample container at 7 A.M.?

18-15 What is the maximum holding time and recommended preservation technique for BOD$_5$ samples?

18-16 The dissolved oxygen meter requires calibration at least once per _____.

18-17 What is the difference between the BOD$_5$ and the CBOD$_5$ test?

18-18 Why is seeding required for samples with high or low pH or chlorinated samples?

18-19 What is the acceptable range of seed correction?

18-20 What is the acceptable preservation method for suspended solids samples?

18-21 Most solids test methods are based upon changes in weight. What can cause changes in weight during the testing procedure?

CHAPTER 19

Permits, Records, and Reports

19.1 INTRODUCTION

PERMITS, records, and reports play a significant role in wastewater treatment operations. In fact, in regards to the "permit," one of the first things any new operator quickly learns is the importance of "making permit" each month. In this chapter, we briefly cover National Pollutant Discharge Elimination System (NPDES) permits and other pertinent records and reports with which the wastewater operator must be familiar.

✓ *Note:* The discussion that follows is general in nature; it does not necessarily apply to any state in particular, but instead is an overview of permits, records, and reports that are an important part of wastewater treatment plant operations. For specific guidance on requirements for your locality, refer to your state water control board or other authorized state agency for information. In this handbook, the term "board" signifies the state reporting agency.

19.2 DEFINITIONS

There are several definitions that should be discussed prior to discussing the permit requirements for records and reporting. These definitions are listed below.

- *Average monthly limitation* the highest allowable average over a calendar month, calculated by adding all of the daily values measured during the month and dividing the sum by the number of daily values measured during the month.
- *Average weekly limitation* the highest allowable average over a calendar week, calculated by adding all of the daily values measured during the calendar week and dividing the sum by the number of daily values determined during the week.
- *Average daily limitation* the highest allowable average over a 24-hour period, calculated by adding all of the values measured during the period and dividing the sum by the number of values determined during the period.
- *Average hourly limitation* the highest allowable average for a 60-minute period, calculated by adding all of the values measured during the period and dividing the sum by the number of values determined during the period.
- *Daily discharge* means the discharge of a pollutant measured during a calendar day or any 24-hour period that reasonably represents the calendar for the purpose of sampling. For pollutants with limitations expressed in units of weight, the daily discharge is calculated as the total mass of the pollutant discharged over the day. For pollutants with limitations expressed in other units, the daily discharge is calculated as the average measurement of the pollutant over the day.
- *Maximum daily discharge* the highest allowable value for a daily discharge.

- *Effluent limitation*   any restriction by the State Board on quantities, discharge rates, or concentrations of pollutants that are discharged from point sources into state waters.
- *Maximum discharge*   the highest allowable value for any single measurement.
- *Minimum discharge*   the lowest allowable value for any single measurement.
- *Point source*   any discernible, defined, and discrete conveyance, including but not limited to any pipe, ditch, channel, tunnel, conduit, well, discrete fissure, container, rolling stock, vessel, or other floating craft, from which pollutants are or may be discharged. This definition does not include return flows from irrigated agricultural land.
- *Discharge monitoring report*   forms for use in reporting of self-monitoring results of the permittee.
- *Discharge permit*   State Pollutant Discharge Elimination System permit that specifies the terms and conditions under which a point source discharge to state waters is permitted.

19.3 NPDES PERMITS

In the United States, all treatment facilities that discharge to state waters must have a discharge permit issued by the State Water Control Board or other appropriate state agency. This permit is known on the national level as the NPDES permit and on the state level as the (state) Pollutant Discharge Elimination System (state-PDES) permit. The permit states the specific conditions that must be met to legally discharge treated wastewater to state waters. The permit contains general requirements (applying to every discharger) and specific requirements (applying only to the point source specified in the permit).

A general permit is a discharge permit that covers a specified class of dischargers. It is developed to allow dischargers with the specified category to discharge under specified conditions.

All discharge permits contain general conditions. These conditions are standard for all dischargers and cover a broad series of requirements. Read the general conditions of the treatment facility's permit carefully.

Permittees must retain certain records. These records include:

- date, time, and exact place of sampling or measurements
- name(s) of the individual(s) performing sampling or measurement
- date(s) and time(s) analyses were performed
- name(s) of the individuals who performed the analyses
- analytical techniques or methods used
- observations, readings, calculations, bench data, and results
- instrument calibration and maintenance
- original strip chart recordings for continuous monitoring
- information used to develop reports required by the permit
- data used to complete the permit application

✓ All records must be kept at least three years (longer at the request of the State Board).

19.3.1 REPORTING

In general (requirements may vary depending upon state regulatory body with reporting authority), reporting must be made under the following conditions/situations:

- unusual or extraordinary discharge reports—Must notify the Board by telephone within 24 hours of occurrence and submit written report within five days. Report must include the following:

    (1) Description of the non-compliance and its cause

(2) Non-compliance date(s), time(s), and duration
(3) Steps planned/taken to reduce/eliminate
(4) Steps planned/taken to prevent recurrence
- anticipated non-compliance—Must notify the Board at least 10 days in advance of any changes to the facility or activity that may result in non-compliance.
- compliance schedules—Must report compliance or non-compliance with any requirements contained in compliance schedules no later than 14 days following scheduled date for completion of the requirement.
- 24-hour reporting—Any non-compliance that may adversely affect state waters or may endanger public health must be reported orally within 24 hours of the time the permittee becomes aware of the condition. A written report must be submitted within five days.
- discharge monitoring reports (DMRs)—Reports self-monitoring data generated during a specified period (normally one month). When completing the DMR, remember the following:
    —More frequent monitoring must be reported.
    —All results must be used to complete reported values.
    —Pollutants monitored by an approved method but not required by the permit must be reported.
    —No empty blocks on the form should be left blank.
    —Averages are arithmetic unless noted otherwise.
    —Appropriate significant figures should be used.
    —All bypasses and overflows must be reported.
    —Report must be signed by the licensed operator.
    —Report must be signed by responsible official.
    —Department must receive by 10th–15th of next month.

19.3.2 SAMPLING AND TESTING

The general requirements of the permit specify minimum sampling and testing that must be performed on the plant discharge. Moreover, the permit will specify the frequency of sampling, sample type, and length of time for composite samples.

Unless a specific method is required by the permit, all sample preservation and analysis must be in compliance with the requirements set forth in the Federal Regulations *Guidelines Establishing Test Procedures for the Analysis of Pollutants under the Clean Water Act* (40 CFR 136).

✓ All samples and measurements must be representative of the nature and quantity of the discharge.

19.3.3 EFFLUENT LIMITATIONS

The permit sets numerical limitations on specific parameters contained in the plant discharge. Limits may be expressed as

- average monthly quantity (kg/day)
- average monthly concentration (mg/L)
- average weekly quantity (kg/day)
- average weekly concentration (mg/L)
- daily quantity (kg/day)
- daily concentration (mg/L)
- hourly average concentration (mg/L)
- instantaneous minimum concentration (mg/L)
- instantaneous maximum concentration (mg/L)

19.3.4 COMPLIANCE SCHEDULES

If the facility requires additional construction or other modifications to fully comply with the final effluent limitations, the permit will contain a schedule of events to be completed to achieve full compliance.

19.3.5 SPECIAL CONDITIONS

Any special requirements or conditions set for approval of the discharge will be contained in this section. Special conditions may include the following:

- monitoring required to determine effluent toxicity
- pretreatment program requirements

19.3.6 LICENSED OPERATOR REQUIREMENTS

The permit will specify, based on the treatment system complexity and the volume of flow treated, the minimum license classification required to be the designated responsible charge operator.

19.3.7 CHLORINATION/DECHLORINATION REPORTING

Several reporting systems apply to chlorination or chlorination followed by dechlorination. It is best to review this section of the specific permit for guidance. If confused, contact the appropriate state regulatory agency.

19.3.8 REPORTING CALCULATIONS

Failure to accurately calculate report data will result in violations of the permit. The basic calculations associated with completing the discharge monitoring report (DMR) are covered below.

19.3.8.1 Average Monthly Concentration

The average monthly concentration is the average of the results of all tests performed during the month.

$$\text{AMC, mg/L} = \frac{\sum \text{Test}_1 + \text{Test}_2 + \text{Test}_3 + \ldots + \text{Test}_n}{N \text{ (Tests during month)}} \qquad (19.1)$$

19.3.8.2 Average Weekly Concentration

The average weekly concentration (AWC) is the average of all the tests performed during a calendar week. A calendar week must start on Sunday and end on Saturday and be completely within the reporting month. A weekly average is not computed for any week that does not meet this criterion.

$$\text{AWC, mg/L} = \frac{\sum \text{Test}_1 + \text{Test}_2 + \text{Test}_3 + \ldots + \text{Test}_n}{N \text{ (Tests during calendar week)}} \qquad (19.2)$$

19.3.8.3 Average Hourly Concentration

The average hourly concentration is the average of all of the test results collected during a 60-minute period.

$$\text{AHC, mg/L} = \frac{\sum \text{Test}_1 + \text{Test}_2 + \text{Test}_3 + \ldots + \text{Test}_n}{N \text{ (Tests during a 60-minute period)}} \quad (19.3)$$

19.3.8.4 Daily Quantity (Kilograms/Day)

Daily quantity is the quantity of a pollutant in kilograms per day discharged during a 24-hour period.

$$\text{Kilograms/Day} = \text{Concentration, mg/L} \times \text{Flow, MGD} \times 3.785 \, \text{kG/MG/mg/L} \quad (19.4)$$

19.3.8.5 Average Monthly Quantity

Average monthly quantity (AMQ) is the average of all the individual daily quantities determined during the month.

$$\text{AMQ, kg/day} = \frac{\sum DQ_1 + DQ_2 + DQ_3 + \ldots + DQ_n}{N \text{ (Tests during month)}} \quad (19.5)$$

19.3.8.6 Average Weekly Quantity

The average weekly quantity is the average of all the daily quantities determined during a calendar week. A calendar week must start on Sunday and end on Saturday and be completely within the reporting month. A weekly average is not computed for any week that does not meet this criterion.

$$\text{AWQ, kg/day} = \frac{\sum DQ_1 + DQ_2 + DQ_3 + \ldots + DQ_n}{N \text{ (Tests during calendar week)}} \quad (19.6)$$

19.3.8.7 Minimum Concentration

The minimum concentration is the lowest instantaneous value recorded during the reporting period.

19.3.8.8 Maximum Concentration

Maximum concentration is the highest instantaneous value recorded during the reporting period.

19.3.8.9 Bacteriological Reporting

Bacteriological reporting is used for reporting fecal coliform test results. To make this calculation, the geometric mean calculation is used, and all monthly geometric means are computed using all the test values. Note that weekly geometric means are computed using the same selection criteria discussed for average weekly concentration and quantity calculations. The easiest method used in making this calculation requires a calculator that can perform logarithmic (log) or Nth root functions.

$$\text{Geometric Mean} = \text{Antilog} \left[\frac{\log X_1 + \log X_2 + \log X_3 + \ldots + \log X_n}{N \text{ (Number of Tests)}} \right] \quad (19.7)$$

or

$$\text{Geometric Mean} = \sqrt[n]{X_1 \times X_2 \times \ldots \times X_n}$$

19.4 CHAPTER REVIEW QUESTIONS

Use the data contained in the chart below:

Date	Day	Flow (MGD)	BOD$_5$ (mg/L)	BOD$_5$ (kg/d)	TSS (mg/L)	TSS (kg/d)	pH (S.U.)	DO (mg/L)
1	F	5.27	10.5	209.4	11.1	221.4	6.5	7.10
2	Sa	4.99					7.2	7.00
3	S	5.40					7.0	7.40
4	M	5.71	27.0	583.5	20.0	432.2	7.2	7.00
5	T	6.46					6.2	7.30
6	W	5.91	9.4	210.3	4.6	102.9	5.2	7.00
7	Th	5.09					5.0	7.00
8	F	5.89	14.3		20.3	452.6	6.0	7.20
9	Sa	5.31					6.6	7.20
10	S	5.82					7.0	7.10
11	M	6.42	4.2	102.1	5.3	128.8	6.1	7.26
12	T	5.72					5.6	7.30
13	W	5.12	21.5	416.7	4.3	83.3	5.5	7.40
14	Th	6.09					5.4	7.30
15	F	5.08	15.7	301.9	11.8	226.9	5.0	7.35
16	Sa	6.22					5.3	7.55
17	S	5.44					5.3	7.10
18	M	5.99	21.7	492.0	17.3	392.2	6.3	7.41
19	T	6.83					5.4	7.25
20	W	5.53	48.3	1,011	75.9	1588.7	4.5	6.88
21	Th	5.38					5.5	7.00
22	F	6.07	73.3	1,684.1	48.7	1118.9	5.2	7.33
23	Sa	5.71					5.5	7.12
24	S	5.40					6.1	7.37
25	M	5.18	13.0	254.9	25.5	500.0	6.2	7.36
26	T	6.09					5.0	6.35
27	W	6.09	17.2	396.5	32.2	742.2	4.9	7.10
28	Th	7.08					5.1	6.40
29	F	5.36	41.8	848.0	55.7	1,130	5.6	7.25
30	Sa	5.48					6.0	7.23
31	S	5.08					6.0	7.17

19-1 What was the daily quantity of BOD$_5$ discharge on day 8?

19-2 What was the monthly average concentration of total suspended solids? If the permit specifies an average concentration of 27 mg/L, how many times did the plant violate its permit limit?

19-3 The plant permit specifies a maximum quantity (weekly average) of 980.4 kg/day. How many times did the plant effluent violate the maximum quantity for BOD$_5$ during the month?

19-4 What was the highest weekly average BOD$_5$ and total suspended solids concentration?

19-5 The permit specifies a minimum dissolved oxygen concentration of 6.0 mg/L. How many times did the effluent violate the permit limit for dissolved oxygen?

19-6 The plant's permit specifies a maximum pH concentration of 9.0 and a minimum concentration of 6.5. How many times did the plant effluent violate the permit limit? What should be recorded in the maximum and minimum columns on the DMR?

19-7 Who must sign the DMR?

19-8 [cross out the wrong word(s)]: If an effluent test is performed more frequently than required by the permit, the extra tests [do/do not] need to be included in the DMR.

CHAPTER 20

Final Review Exam

20.1 INTRODUCTION

Now that you have reviewed each lesson and have completed the chapter review questions, you may test your overall knowledge of the material contained in Volume 1 of the handbook by completing the following review examination. For the questions you have difficulty answering or you answer wrong, review the pertinent sections containing the applicable subject matter. Successful review and completion of all the requirements specified in this edition of the handbook should prepare you for the Class IV/Grade I licensing examinations—and should set the stage for successful completion of the Class III/Grade II examinations. For those operators preparing for the Class III/Grade II and Class II/Grade III licensing exams, review and completion of the requirements in Volume 2 of the handbook (Intermediate Level) is highly recommended.

Unlike the actual state licensure examinations, which contain an assortment of different types of questions (i.e., multiple choice, true or false, essay, completion questions, etc.), the final review examinations presented in each handbook require a written response to each question. I have formatted the examinations this way because experience has shown me that when studying for an exam (any exam), it is always best to write out the "correct" answer (for retention purposes). Moreover, when studying for an exam, it is best to view only the correct answer instead of several different choices that might be confused as being the correct answer—which could enable the test taker to select the wrong answer on the licensure exam.

Upon completion of the final review exam, check your answers with those given in Appendix B.

20.2 REVIEW EXAM

20-1 Give three reasons for treating wastewater.

20-2 Name two types of solids based on physical characteristics.

20-3 Define organic and inorganic.

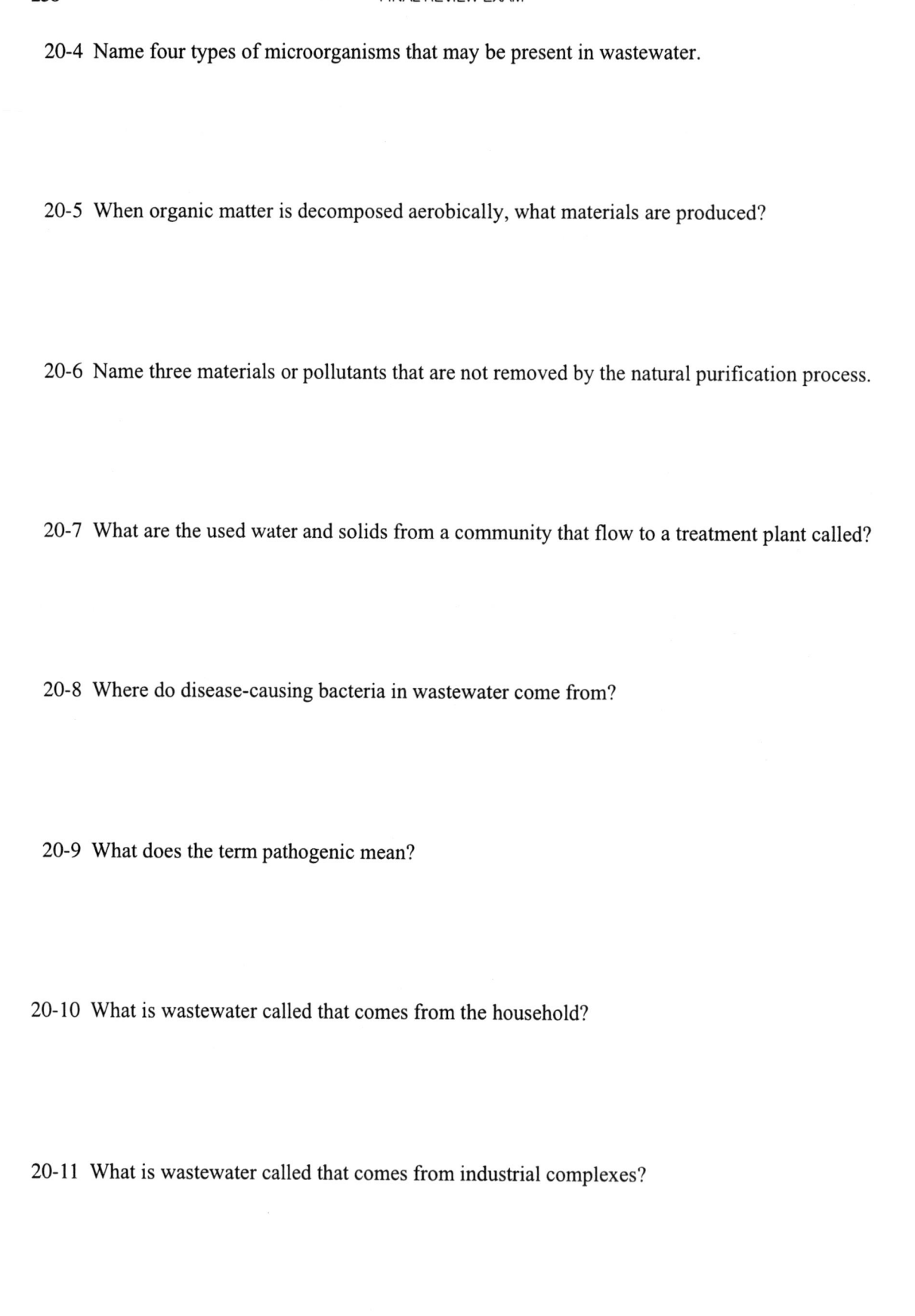

20-4 Name four types of microorganisms that may be present in wastewater.

20-5 When organic matter is decomposed aerobically, what materials are produced?

20-6 Name three materials or pollutants that are not removed by the natural purification process.

20-7 What are the used water and solids from a community that flow to a treatment plant called?

20-8 Where do disease-causing bacteria in wastewater come from?

20-9 What does the term pathogenic mean?

20-10 What is wastewater called that comes from the household?

20-11 What is wastewater called that comes from industrial complexes?

The following information is used for questions 20-12 through 20-17:

Primary settling	Number	2
	Length	180 ft
	Width	110 ft
	Water Depth	12 ft
Aeration tank	Number	4
	Length	210 ft
	Width	90 ft
	Water Depth	14 ft
Secondary settling tank	Number	4
	Diameter	120 ft
	Water Depth	18 ft

20-12 The effluent weir on the secondary settling tank is located along the outer edge of the tank. What is the weir length in feet for each settling tank?

20-13 What is the surface area of each of the primary settling tanks in square feet?

20-14 What is the volume of each of the aeration tanks in cubic feet?

20-15 You wish to install a fence around each aeration tank to prevent falls into the tanks. How many feet of fence must be ordered?

20-16 The secondary settling tanks consist of a cylindrical section 18 ft deep and a cone-shaped bottom that has a depth of 8 ft. What is the total volume of each settling tank in cubic feet?

20-17 The lab test indicates that a 500-g sample of sludge contains 22 g of solids. What is the percent solids in the sludge sample?

20-18 The depth of water in the grit channel is 28 in. What is the depth in feet?

20-19 The operator withdraws 5,250 gal of solids from the digester. How many pounds of solids have been removed?

20-20 Sludge added to the digester causes a 1,920 cubic foot change in the volume of sludge in the digester. How many gallons of sludge have been added?

20-21 The plant effluent contains 30 mg/L of solids. The effluent flow rate is 3.40 MGD. How many pounds per day of solids are discharged?

20-22 The plant effluent contains 25 mg/L BOD_5. The effluent flow rate is 7.25 MGD. How many kilograms per day of BOD_5 are being discharged?

20-23 The operator wishes to remove 3,280 lb per day of solids from the activated sludge process. The waste-activated sludge concentration is 3,250 mg/L. What is the required flow rate in million gallons per day?

20-24 The plant influent includes an industrial flow that contains 240 mg/L BOD. The industrial flow is 0.72 MGD. What is the population equivalent for the industrial contribution in people per day?

20-25 The label of hypochlorite solution states that the specific gravity of the solution is 1.1288. What is the weight of 1 gal of the hypochlorite solution?

The following information is used for questions 20-26 through 20-29:

Plant influent	Flow	8.40 MGD
	Suspended solids	370 mg/L
	BOD	230 mg/L
Primary effluent	Flow	8.40 MGD
	Suspended solids	130 mg/L
	BOD	170 mg/L
Activated sludge eff.	Flow	8.34 mg/L
	Suspended solids	18 mg/L
	BOD	22 mg/L
Anaerobic digester	Solids in	6.6%
	Solids out	13.5%
	Volatile matter in	66.5%
	Volatile matter out	48.9%

20-26 What is the plant percent removal for BOD_5?

20-27 What is the plant percent removal for TSS?

20-28 What is the primary treatment percent removal for BOD_5?

20-29 What is the percent volatile matter reduction in the digestion process?

The following information is used for questions 20-30 through 20-32:

Plant influent	Flow	8.40 MGD
Grit channel	Number of channels	2
	Channel length	40 ft
	Channel width	3 ft
	Water depth	2.5 ft
Primary settling	Number	2
	Length	140 ft
	Width	100 ft
	Water depth	12 ft
Anaerobic digester	Flow	19,000 gpd
	Volume	115,000 ft^3

20-30 What is the hydraulic detention time in hours for primary settling when both tanks are in service?

20-31 What is the hydraulic detention time in the grit channel in minutes when both channels are in service?

20-32 What is the hydraulic detention time of the anaerobic digester in days?

20-33 What is the purpose of the bar screen?

20-34 What two methods are available for cleaning a bar screen?

20-35 Name two ways to dispose of screenings.

20-36 What must be done to the cutters in a comminutor to ensure proper operation?

20-37 The comminutor jams frequently. A review of the maintenance records indicates that the cutters were changed approximately 2 weeks ago and the cutter alignment was checked yesterday. What are the possible causes for the continued jamming problem? What actions would you recommend to identify the specific cause?

20-38 What is grit? Give three examples of material that is considered to be grit.

20-39 The plant has three channels in service. Each channel is 2 ft wide and has a water depth of 3 ft. What is the velocity (fps) in the channel when the flow rate is 8.0 MGD?

20-40 The grit from the aerated grit channel has a strong hydrogen sulfide odor upon standing in a storage container. What does this indicate, and what action should be taken to correct the problem?

20-41 What is the purpose of primary treatment?

20-42 What is the purpose of the settling tank in the secondary or biological treatment process?

20-43 The circular settling tank is 90 ft in diameter and has a depth of 12 ft. The effluent weir extends around the circumference of the tank. The flow rate is 2.25 MGD. What is the detention time in hours, surface loading rate in gal/day/ft^2, and weir overflow rate in gal/day/ft?

The following information is used for questions 20-44 through 20-51:

Plant effluent	Flow	0.44 MGD
	Suspended solids	370 mg/L
	BOD	380 mg/L
Town population	People	2,850 people
Industrial contribution	Flow	0.039 MGD
	Suspended solids	650 mg/L
	BOD	970 mg/L
Pond	Length	1,500 ft
	Width	1,100 ft
	Operating depth	4.1 ft

20-44 What is the pond area in acres?

20-45 What is the pond volume in acre-feet?

20-46 What is the influent flow rate in acre-feet per day?

20-47 What is the influent flow rate in acre-inches per day?

20-48 What is the pond hydraulic detention time in days?

20-49 What is the pond hydraulic loading in inches/day?

20-50 What is the pond organic loading in pounds of BOD per acre per day?

20-51 What is the current population loading (including the industrial contributions)?

20-52 Give three classifications of ponds based upon their location in the treatment system.

20-53 Describe the processes occurring in a raw sewage stabilization pond (facultative).

20-54 How do changes in the season affect the quality of the discharge from a stabilization pond?

20-55 What is the advantage of using mechanical or diffused aeration equipment to provide oxygen?

20-56 Describe how the dissolved oxygen level of the pond changes during the day.

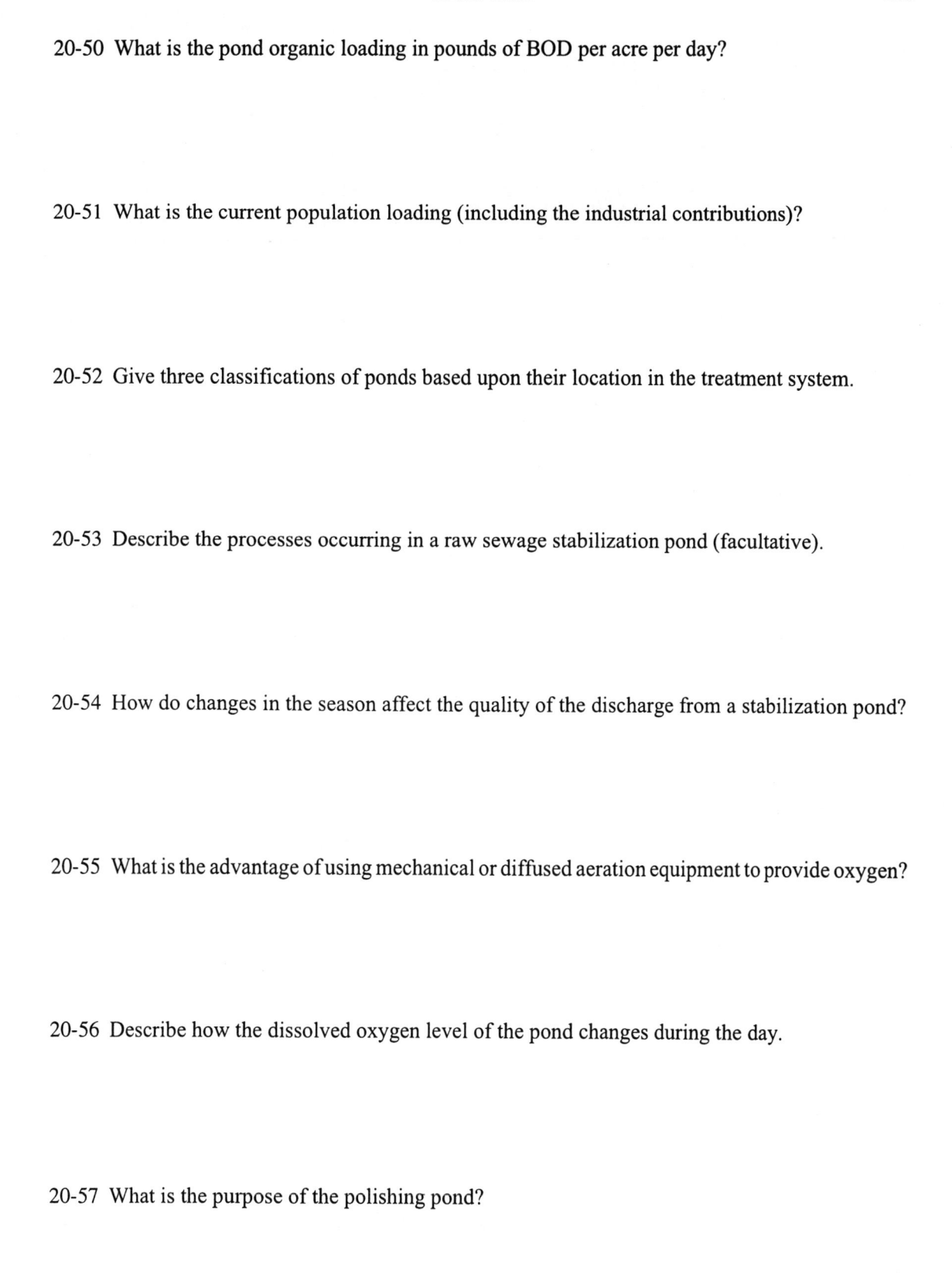

20-57 What is the purpose of the polishing pond?

The following information is used for questions 20-58 through 20-62:

Plant effluent	Flow	2.40 MGD
	Suspended solids	205 mg/L
	BOD_5	196 mg/L
Trickling filters	Number	2
	Diameter	90 ft
	Media depth	8 ft
Recirculation	Ratio	1.5:1.0

20-58 What is the total flow to each filter in million gallons per day? (Assume the flow is equally split.)

20-59 What will the total flow to each filter be in million gallons per day if the operator changes the recirculation rate to 0.7:1.0? (Assume the flow is equally split between the two filters.)

20-60 What is the hydraulic loading in gallons per day per square foot for each filter at a 1.5:1.0 recirculation rate? (Assume the flow is equally split between the two filters.)

20-61 What is the hydraulic loading in gallons per day per square foot for each filter at a 0.75:1.0 recirculation rate? (Assume the flow is equally split between the two filters.)

20-62 What is the organic loading in pounds of BOD per 1,000 ft^3 for each filter at a 0.80:1.0 recirculation rate? (Assume the flow is equally split between the filters.)

20-63 Name three main parts of the trickling filter, and give the purposes of each part.

20-64 Name three classifications of trickling filters, and identify the classification that produces the highest quality effluent.

The following information is used for questions 20-65 through 20-67:

RBC influent	Flow	8.40 MGD
	Suspended solids	150 mg/L
	BOD$_5$	190 mg/L
RBC design	Number of stages	8
	Media area—Stage 1	250,000 ft^2
	Media area—Stage 2	220,000 ft^2
	Media area—Stage 3, 4, 5, 6, 7, 8	150,000 ft^2 (each)
	K factor	0.55
	Hydraulic loading	7.1 gpd/ft^2
	SOL	6.5 lbs BOD/1,000 ft^2

20-65 What is the hydraulic loading in gal/day/ft^2?

20-66 What is the total organic loading (TOL) in pounds of BOD/day/1,000 ft^2 of media?

20-67 What is the soluble BOD organic loading (SOL) in pounds of BOD/day/1,000 ft^2 of media?

20-68 Describe the process occurring in the rotating biological contactor process.

20-69 What makes the RBC process similar to the trickling filter?

20-70 Describe the appearance of the slime when the RBC is operating properly. What happens if the RBC is exposed to a wastewater containing high amounts of sulfur?

The following information is used for questions 20-71 through 20-75:

Influent		Flow	2.10 MGD
		BOD	230 mg/L
		TSS	370 mg/L
Primary effluent		Flow	2.10 MGD
		BOD	175 mg/L
		TSS	130 mg/L
Activated sludge process effluent		Flow	2.10 MGD
		BOD	22 mg/L
		TSS	18 mg/L
Aeration tank		Volume	1,255,000 gal
		MLSS	2,750 mg/L
		MLVSS	1,850 mg/L
SSV test		Sample	2,000 mL
		30 minutes	1,750 mL
		60 minutes	1,050 mL
Waste		Flow	0.090 MGD
		Solids	6,120 mg/L
		Volatile	66%
Return		Flow	0.50 MGD
		Solids	5,750 mg/L
Settling tank		Volume	880,000 gal
Desired F/M ratio			0.25 lb/lb
Desired MCRT			6.5 days

20-71 What is the percent SSV_{30}?

20-72 What is the SSV_{60} mL/L?

20-73 Using the SSV_{30} in mL/L, what is the sludge volume index?

20-74 What is the food to microorganism (F/M) ratio?

20-75 What is the mean cell residence time in days? (Assume the clarifier solids concentration equals the aeration tank MLSS.)

20-76 Microscopic examination reveals a predominance of rotifers. What process adjustment does this indicate is required?

20-77 Increasing the wasting rate will _____ the MLSS, _____ the return concentration, _____ the MCRT, _____ the F/M ratio, and _____ the SVI.

20-78 The plant adds 320 lb/day of dry hypochlorite powder to the plant effluent. The hypochlorite powder is 45% available chlorine. What is the chlorine feed rate in pounds per day?

20-79 The plant uses liquid HTH, which is 67.9% available chlorine and has a specific gravity of 1.18. The required feed rate to comply with the plant's discharge permit total residual chlorine limit is 285 lb/day. What is the required flow rate for HTH solution in gallons per day?

20-80 The plant currently uses 45.8 lb of chlorine per day. Assuming the chlorine usage will increase by 10% during the next year, how many 2,000-lb cylinders of chlorine will be needed for the year (365 days)?

20-81 The plant has six 2,000-lb cylinders on hand. The current dose of chlorine being used to disinfect the effluent is 6.2 mg/L. The average effluent flow rate is 2.25 MGD. Allowing 15 days for ordering and shipment, when should the next order for chlorine be made?

20-82 The plant feeds 38 lb of chlorine per day and uses 150-lb cylinders. Chlorine use is expected to increase by 11% next year. The chlorine supplier has stated that the current price of chlorine ($0.170 per pound) will increase by 7.5% next year. How much money should the town budget for chlorine purchases for the next year (365 days)?

20-83 What is the difference between disinfection and sterilization?

20-84 To be effective, chlorine must be added to satisfy the _____ and produce a _____ mg/L _____ for at least _____ minutes at design flow rates.

20-85 Elemental chlorine gas is _____ in color and is _____ times heavier than air.

20-86 The sludge pump operates 30 minutes every three hours. The pump delivers 70 gpm. If the sludge is 5.1% solids and has a volatile matter content of 66%, how many pounds of volatile solids are removed from the settling tank each day?

20-87 The aerobic digester has a volume of 63,000 gal. The laboratory test indicates that 41 mg of lime were required to increase the pH of a 1-L sample of digesting sludge from 6.0 to the desired 7.1. How many pounds of lime must be added to the digester to increase the pH of the unit to 7.1?

20-88 The digester has a volume of 73,500 gal. Sludge is added to the digester at the rate of 2,750 gal/day. What is the sludge retention time in days?

20-89 What is the normal operating temperature of a heated anaerobic digester? What is the maximum change that should be made in a day to avoid reductions in gas production?

20-90 The supernatant contains 340 mg/L volatile acids and 1,830 mg/L alkalinity. What is the volatile acids alkalinity ratio?

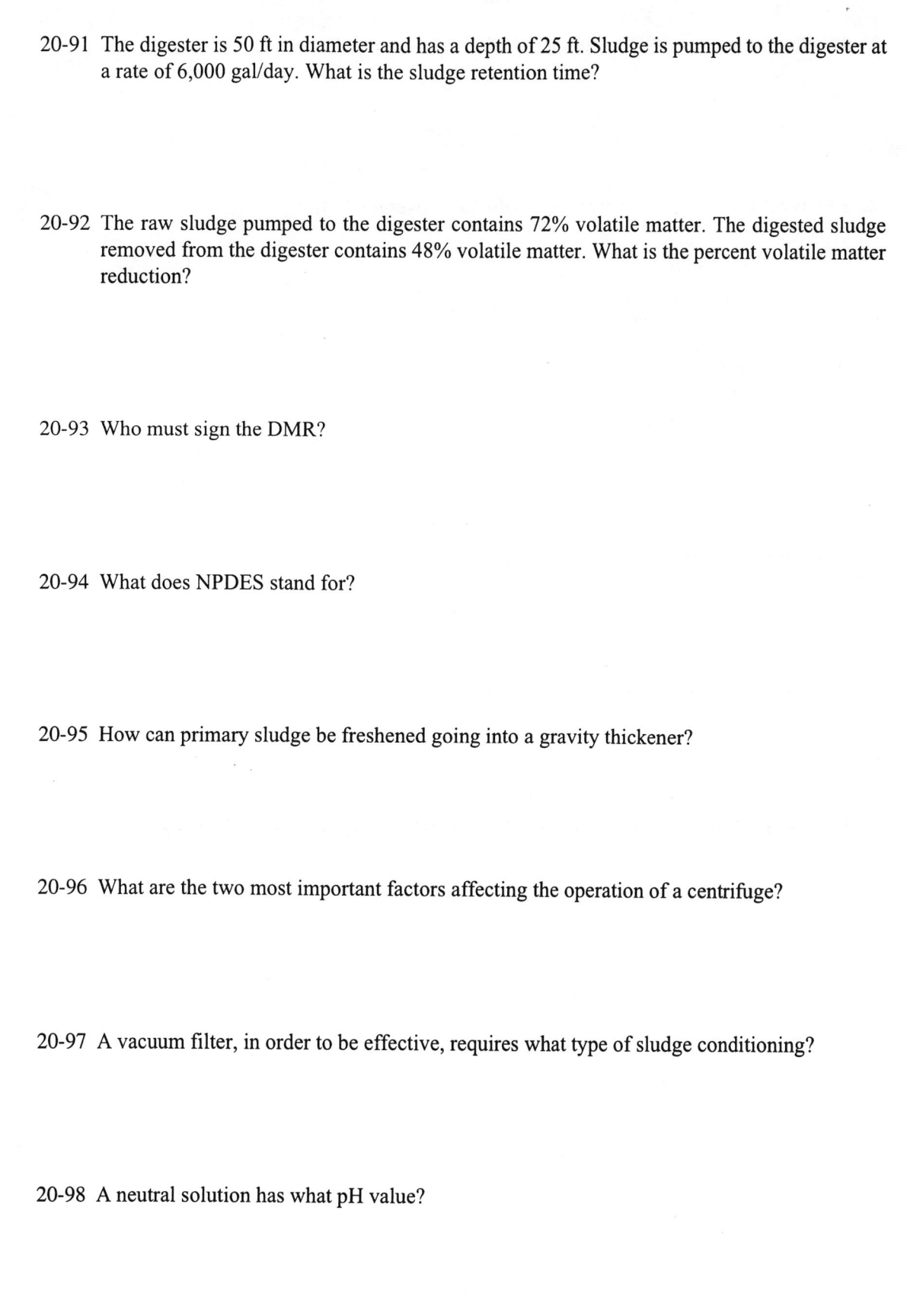

20-91 The digester is 50 ft in diameter and has a depth of 25 ft. Sludge is pumped to the digester at a rate of 6,000 gal/day. What is the sludge retention time?

20-92 The raw sludge pumped to the digester contains 72% volatile matter. The digested sludge removed from the digester contains 48% volatile matter. What is the percent volatile matter reduction?

20-93 Who must sign the DMR?

20-94 What does NPDES stand for?

20-95 How can primary sludge be freshened going into a gravity thickener?

20-96 What are the two most important factors affecting the operation of a centrifuge?

20-97 A vacuum filter, in order to be effective, requires what type of sludge conditioning?

20-98 A neutral solution has what pH value?

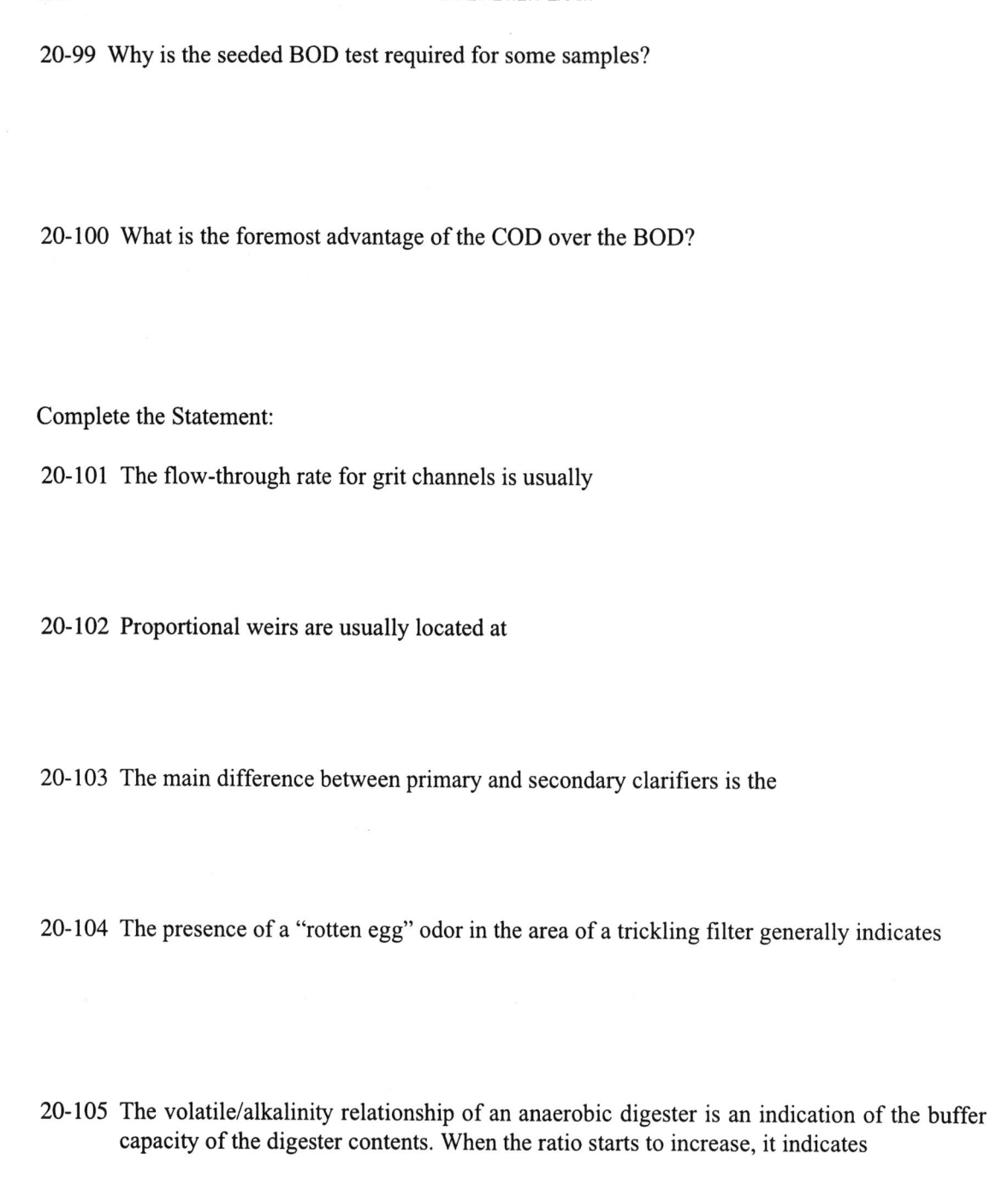

20-99 Why is the seeded BOD test required for some samples?

20-100 What is the foremost advantage of the COD over the BOD?

Complete the Statement:

20-101 The flow-through rate for grit channels is usually

20-102 Proportional weirs are usually located at

20-103 The main difference between primary and secondary clarifiers is the

20-104 The presence of a "rotten egg" odor in the area of a trickling filter generally indicates

20-105 The volatile/alkalinity relationship of an anaerobic digester is an indication of the buffer capacity of the digester contents. When the ratio starts to increase, it indicates

APPENDIX A

Answers to Chapter Review Questions

Exercise 3.1

3.1 24
3.2 27
3.3 18
3.4 10
3.5 20

Chapter 3 Review Questions:

3-1   0

3-2 $C = D\pi$ $140' \times 3.14 = 440$ ft

3-3 $A = LD$ $120' \times 60' = 7{,}200$ ft^2

3-4 $20' + 5'$ $D = 80'$

0.785×80 ft $\times 80$ ft $\times 20$ ft $= 100{,}480$ ft^3

0.262×80 ft $\times 80$ ft $\times 5$ ft $= \underline{8{,}384}$ ft^3

$108{,}864$ ft^3

3-5 $\dfrac{6.5\%}{100} \times 8{,}000 \times 8.34 = 4{,}337$ lb

3-6 $\dfrac{20 \text{ g} \times 100}{600 \text{ g}} = 3.3\%$

3-7 $\dfrac{20 \text{ g} \times 100}{500 \text{ g}} = 4\%$

3-8 3.14×100 ft $= 314$ ft

3-9 130 ft $\times 110$ ft $= 14{,}300$ ft^2

3-10 250×110 ft $\times 14$ ft $= 385{,}000$ ft^3

3-11 $(2 \times 250$ ft$) + (2 \times 110$ ft$) = 720$ ft $\times 4 = 2{,}880$ ft

3-12 0.785×100 ft $\times 100$ ft $\times 18$ ft $= 141{,}300$ ft^3

0.262×100 ft $\times 100$ ft $\times 10$ ft $= \underline{26{,}200}$ ft^3

$167{,}500$ ft^3

3-13 32 mg/L

253

3-14 Day 8   34.7 mg/L
      Day 9   31.0 mg/L
3-15 Antilog of Geometric Mean = 1.39967
      Geometric Mean = 25.1 mpn/100 mL

Chapter 4 Review Questions:

4-1 $\dfrac{36 \text{ in.}}{12 \text{ in./ft}} = 3 \text{ ft}$

4-2 $5{,}269 \text{ gal} \times 8.34 \text{ lb/gal} = 43{,}943 \text{ lb}$

4-3 $1{,}996 \text{ ft}^3 \times 7.48 \text{ gal/ft}^3 = 14{,}930 \text{ gal}$

4-4 $35 \text{ mg/L} \times 3.69 \text{ MGD} \times 8.34 \text{ lb/MG/mg/L} = 1{,}077 \text{ lb/day}$

4-5 $26 \text{ mg/L} \times 7.25 \text{ MGD} \times 3.785 \text{ KG/MG/mg/L} = 713 \text{ KG/day}$

4-6 $\dfrac{3{,}540 \text{ lb/day}}{3{,}524 \text{ mg/L} \times 8.34 \text{ lb/MG/L}} = 0.120 \text{ MG}$

4-7 $\dfrac{235 \text{ mg/L} \times 0.70 \text{ MGD} \times 8.34 \text{ lb/MG/mg/L}}{0.17 \text{ lb BOD}_5/\text{person}} = 8{,}070 \text{ people}$

4-8 BOD = 0.21 lb/cap · d
     SS = 0.28 lb/cap · d

   Per Capita BOD, mg/L Concentration =

   $\dfrac{0.21 \text{ lb/capita} \cdot \text{d} \times 10^6 \text{ gal/M gal}}{[8.34 \text{ lb/M gal} \cdot (\text{mg/L})] \times 110 \text{ gal/capita} \cdot \text{d}} = 229 \text{ mg/L}$

   Per Capita SS, mg/L Concentration =

   $\dfrac{0.28 \text{ lb/capita} \cdot \text{d} \times 10^6 \text{ gal/M gal}}{[8.34 \text{ lb/M gal} \cdot (\text{mg/L})] \times 110 \text{ gal/capita} \cdot \text{d}} = 305 \text{ mg/L}$

4-9 $8.34 \text{ lb/gal} \times 1.1545 = 9.63 \text{ lb/gal}$

Chapter 5 Review Questions:

5-1 $\dfrac{(225 \text{ mg/L} - 24 \text{ mg/L}) \times 100}{225 \text{ mg/L}} = 89\%$

5-2 $\dfrac{(350 \text{ mg/L} - 17 \text{ mg/L}) \times 100}{350 \text{ mg/L}} = 95\%$

5-3 $\dfrac{(225 \text{ mg/L} - 175 \text{ mg/L}) \times 100}{225 \text{ mg/L}} = 22\%$

5-4 $\dfrac{(350 \text{ mg/L} - 144 \text{ mg/L}) \times 100}{350 \text{ mg/L}} = 59\%$

5-5 $\dfrac{(0.663 - 0.491) \times 100}{[0.663 - (0.663 \times 0.491)]} = 51\%$

Chapter 6 Review Questions:

6-1 Tank Volume $= 70 \text{ ft} \times 16 \text{ ft} \times 10 \text{ ft} \times \dfrac{7.48 \text{ gal}}{\text{ft}^3} = 83,776 \text{ gal}$

$1.35 \text{ MGD} = \dfrac{1,350,000}{1 \text{ d}}$

Detention Time $= 83,776 \text{ gal} \times \dfrac{1 \text{ day}}{1,350,000 \text{ gal}} \times \dfrac{24 \text{ hr}}{1 \text{ d}} = 1.5 \text{ hr}$

6-2 $\dfrac{160 \text{ ft} \times 110 \text{ ft} \times 12 \text{ ft} \times 2 \text{ tanks} \times 7.48 \text{ gal}/\text{ft}^3 \times 24 \text{ h/day}}{8.40 \text{ MGD} \times 1,000,000 \text{ gal}/\text{MG}} = 9 \text{ h}$

6-3 $\dfrac{60 \text{ ft} \times 4 \text{ ft} \times 2.6 \text{ ft} \times 2 \text{ channels} \times 7.48 \text{ gal}/\text{ft}^3 \times 1,440 \text{ min}/\text{day}}{8.40 \text{ MGD} \times 1,000,000 \text{ gal}/\text{MG}} = 1.6 \text{ min}$

6-4 $\dfrac{110,000 \text{ ft}^3 \times 7.48 \text{ gal}/\text{ft}^3}{19,000 \text{ gpd}} = 43 \text{ days}$

Chapter 7 Review Questions:

7-1 The amount of organic matter that can be biologically oxidized under controlled conditions (5 days @ 20°C in the dark).

7-2 Human and animal wastes—body discharges
Household waste—garbage, paper, cleaning material, etc.
Industrial waste—waste material from industrial processes

7-3 Organic indicates matter that is made up mainly of carbon, hydrogen, and oxygen and will decompose into mainly carbon dioxide and water at 550°C.
Inorganic indicates mineral matter that is made up of elements such as aluminum, iron, sodium, chlorine, etc. (substances that will not be destroyed when burned at 550°C).

7-4 Dissolved and suspended.

7-5 Water from street drains, parking lots, roof drains, etc. It can hydraulically overload the plant.

7-6 Domestic, sanitary, and industrial.

7-7 Prevent disease; protect aquatic organisms; protect water quality

Chapter 8 Review Questions:

8-1 Algae, bacteria, rotifers, viruses, protozoa.

8-2 Pathogenic organisms.

8-3 Carbon dioxide, stable solids, more organisms.

8-4 Color turns gray, solids settle, DO decreases rapidly, fish disappear, microorganism population increases rapidly.

8-5 Zone of active decomposition.

8-6 Process can return to degradation or active decomposition zones.

8-7 Inorganic solids; toxic materials; pathogenic organisms.

8-8 Proper pH; adequate supply of organic matter (biodegradable); adequate supply of oxygen; enough nutrients, not toxic matter, enough water.

Chapter 9 Review Questions:

9-1 $669.9 \text{ ft}^3 \times 7.48 \text{ gal/ft}^3 = 5,011 \text{ gal}$

9-2 (1) $120 \text{ lb} - 85 \text{ lb} = 35$
     (2) specific gravity $= 120/35 = 3.4$

9-3 (1) $(8.34 \text{ lb/gal})(0.91) = 7.59 \text{ lb/gal}$
     (2) $(1,270 \text{ gal})(7.59 \text{ lb/gal}) = 9,639 \text{ lb}$

9-4 $65 \text{ ft} \times 0.433 \text{ psi/ft} = 28.1 \text{ ft}$

9-5 Height in feet $= \dfrac{14 \text{ psi}}{0.433 \text{ psi/ft}} = 32.3 \text{ ft}$

9-6 Static head = Discharge elevation − Supply elevation
     $2,566 - 2,133 \text{ ft} = 433 \text{ ft}$

Chapter 10 Review Questions:

10-1 $2,225 \text{ ft} - 1,810 \text{ ft} = 415 \text{ ft}$

$415 \text{ ft} + 310 \text{ ft} + 175 \text{ ft} = 900 \text{ ft}$ total head

Water Horsepower $= \dfrac{425 \text{ gpm} \times 900 \text{ ft} \times 8.34 \text{ lb/gal}}{33,000 \text{ ft-lb/hp}} = 96.7 \text{ whp}$

Brake Horsepower $= \dfrac{96.7 \text{ whp}}{0.71} = 136.2 \text{ bhp}$

Motor Horsepower $= \dfrac{136.2 \text{ bhp}}{0.96} = 141.9 \text{ mhp}$

10-2 $141.9 \text{ hp} \times 0.746 \text{ kW/hp} = 105.9 \text{ kW}$

10-3 $150 \text{ hp} \times 0.746 \text{ kW/hp} \times 12 \text{ hours} \times 5 \text{ days} \times 52 \text{ weeks} \times \$0.02333 = \$8,145$

Chapter 11 Review Question:

11-1 To lift the wastewater to provide energy for continued flow to the treatment plant.

Chapter 12 Review Questions:

12-1 To protect plant equipment and remove materials that are not affected by treatment.
12-2 To remove large solids (rags, sticks, rocks, etc.).
12-3 Manual, mechanical.
12-4 Burial, incineration, grinding, and return to flow.
12-5 Sharpen and align.
12-6 The rate of flow, the depth of the wastewater in the channel, the width of the channel, and number of channels in service.
12-7 $V, \text{fps} = \dfrac{8.0 \text{ MGD} \times 1.55 \text{ cfs/MGD}}{3 \text{ channels} \times (2 \text{ ft}) \times (3 \text{ ft})} = .69 \text{ fps}$

12-8 Reduce odors, freshen septic wastes, reduce BOD_5, prevent corrosion, improve settling and flotation.

12-9 To prevent excessive wear grit causes to pumps and to prevent it from taking up valuable space in downstream units.

12-10 1 ft/s

12-11 1.4 ft/s

12-12 $\dfrac{4.5\,\text{MG}}{1\,\text{d}} \times \dfrac{1\,\text{d}}{24\,\text{h}} \times \dfrac{84\,\text{h}}{1} = 15.75\,\text{MG}$

$\dfrac{4\,\text{ft} \times 5\,\text{ft} \times 2\,\text{ft}}{15.75\,\text{MG}} = 2.5\,\text{ft}^3/\text{MG}$

12-13 Anaerobic.

12-14 Flow rate in open channels.

12-15 Decrease slowly, then return to saturation slowly.

12-16 Slow the wastewater so that the heavy inorganic material will settle out.

12-17 Remove settleable solids.

Chapter 13 Review Questions:

13-1 To reduce settleable and flotable solids.

13-2 $\text{VM lb/day} = 75\,\text{gpm} \times \dfrac{20\,\text{min}}{3\,\text{h}} \times \dfrac{24\,\text{h}}{\text{day}} \times \dfrac{5.5\%}{100} \times \dfrac{66\%}{100} \times \dfrac{8.34\,\text{lb}}{\text{gal}} = 3{,}633\,\text{lb/day}$

13-3 $\dfrac{0.785 \times 80\,\text{ft} \times 80\,\text{ft} \times 12\,\text{ft} \times 7.48\,\text{gal/ft}^3 \times 24\,\text{h/day}}{2.6\,\text{MGD} \times 1{,}000{,}000\,\text{gal/MG}} = 4.2\,\text{h}$

✓ *Note:* 1–3 hours is recommended; obviously, 4.2 hours exceeds this limit.

$\dfrac{2.6\,\text{MGD} \times 1{,}000{,}000\,\text{gal/MG}}{0.785 \times 80\,\text{ft} \times 80\,\text{ft}} = 517.5\,\text{gpd/ft}^2$

$\dfrac{2.6\,\text{MGD} \times 1{,}000{,}000\,\text{gal/MG}}{3.14 \times 80\,\text{ft}} = 10{,}350\,\text{gpd/ft}$

13-4 Operate sludge pumping often enough to prevent large amounts of septic solids on the surface of the settling tank while maintaining a % sludge solids of 4–8%.

13-5 To prevent flotable solids (scum) from leaving the tank.

13-6 90–95% of the settleable solids.

13-7 1.5 to 2.5 hours.

13-8 Surface = $80\,\text{ft} \times 20\,\text{ft} = 1600\,\text{ft}^2$

Surface overflow rate = $\dfrac{1{,}500{,}000\,\text{gpd}}{1{,}600\,\text{ft}^2} = 937.5\,\text{gpd/ft}^2$

13-9 $\text{WOR} = \dfrac{1{,}250{,}000\,\text{gpd}}{80\,\text{ft}} = 15{,}625\,\text{gpd/ft}$

13-10 Tank Vol = $80\,\text{ft} \times 20\,\text{ft} \times 12\,\text{ft} \times 7.48\,\text{gal/ft}^3 = 143{,}616\,\text{gal}$

Detention Time = $3 \times 143{,}616\,\text{gal} \times \dfrac{1\,\text{d}}{5{,}000{,}000} \times \dfrac{24\,\text{h}}{1\,\text{d}} \times \dfrac{60\,\text{min}}{1\,\text{h}} = 124\,\text{min}$

$\text{SOR} = \dfrac{5{,}000{,}000\,\text{gpd}}{3 \times 80\,\text{ft} \times 20\,\text{ft}} = 1{,}042\,\text{gpd/ft}^2$

$$\text{WOR} = \frac{5,000,000 \text{ gpd}}{3 \times 86 \text{ ft}} = 19,380 \text{ gpd/ft}$$

13-11   2.8 hours.

Chapter 14 Review Questions:

14-1   Facultative ponds.

14-2   Stabilization pond, oxidation pond, polishing pond.

14-3   Settling, anaerobic digestion of settled solids, aerobic/anaerobic decomposition of dissolved and colloidal organic solids by bacteria producing stable solids and carbon dioxide, photosynthesis production of oxygen by algae.

14-4   Summer effluent is high in solids (algae) and low in BOD_5.
Winter effluent is low in solids and high in BOD_5.

14-5 $609 \text{ ft} \times 425 \text{ ft} \times 6 \text{ ft} \times \dfrac{7.48 \text{ gal}}{1 \text{ ft}^3} = 13,161,060 \text{ gal}$

$13,161,060 \text{ gal} \times \dfrac{1 \text{ d}}{300,000 \text{ gal}} = 43.9 \text{ d}$

14-6 $730 \text{ ft} \times 410 \text{ ft} \times \dfrac{1 \text{ ac}}{43,560 \text{ ft}^3} = 6.87 \text{ ac}$

$\dfrac{0.66 \text{ ac-ft/d}}{6.87 \text{ ac}} = 1.15 \text{ in./d}$

14-7   (1) Distribution—to spread the wastewater evenly over the media.
(2) Media—to support the biological growth.
(3) Underdrains—to collect the flow and transport it out of the filter, provide ventilation, and support the media.

14-8   Standard rate, high rate, roughing.

14-9   Standard rate.

14-10   To provide additional oxygen, reduce organic loading, improve sloughing, reduce odors, eliminate filter flies.

14-11   Total = 2.3 MGD × (1.0 + 0.80) = 4.1 MGD

14-12   To remove sloughings from the wastewater prior to discharge.

14-13   Arm movement, distribution, orifice clogging, odors, operational problem indications.

14-14   0.288 MGD + 0.366 MGD = 0.654 MGD

Surface Area = $0.785 \times (90 \text{ ft})^2 = 6,359 \text{ ft}^2$

$\dfrac{654,000}{6,359 \text{ ft}^2} = 103 \text{ gpd/ft}^2$

14-15 $\dfrac{4.30 \text{ MGD}}{3.0 \text{ MGD}} = 1.43$

14-16   A series of plastic disks placed side by side on a shaft. The disks are suspended in a channel of wastewater and rotate through the wastewater.

14-17   Slime on the disks collects organic solids from the wastewater; organisms biologically oxidize the materials to produce stable solids. As the disk moves through the air, oxygen is transferred to the slime to keep it aerobic. Excess solids are removed as sloughings as the disk moves through the wastewater.

Answers to Chapter Review Questions

14-18 No. An RBC must follow primary settling in order to remove the settleable solids. An RBC is designed to treat only soluble material.

14-19 White biomass indicates filamentous bacteria growth.

14-20 Standard density and high density.

14-21 The biological growth is attached to the media.

14-22 The units are normally covered and maintain the same temperature throughout the year.

14-23 Gray, shaggy, translucent is normal. Sulfur will cause slime to become white and chalky in appearance.

14-24 Nitrification is occurring in the later stages of the process.

14-25 $\dfrac{450,000 \text{ gpd}}{200,000 \text{ ft}^2} = 2.25 \text{ gpd/ft}^2$

Chapter 15 Review Questions:

15-1 Provides the needed oxygen for the microorganisms and the required mixing to put the food and microorganisms together.

15-2 Activated sludge.

15-3 Mixed liquor.

15-4 Food, oxygen, and organisms.

15-5 To remove BOD.

15-6 Mechanical and diffused.

15-7 Aeration tank color, foam, odors, settling tank capacity, solids loss, aeration rates, process control tests, etc.

15-8 $SSV = \dfrac{1,300 \text{ mL} \times 1,000}{2,200 \text{ mL}} = 591$

15-9 $\dfrac{420 \times 1,000 \text{ gpd}}{2,245 \text{ mL}} = 187$

15-10 waste lb/day = $8,185 \text{ mg/L} \times 0.069 \text{ MGD} \times 8.34 = 4,710$

15-11 Contact stabilization.

15-12 Gravity thickening, dissolved air flotation thickening, and sludge concentration (belt thickener).

Chapter 16 Review Questions:

16-1 Disinfection destroys pathogenic organisms. Sterilization destroys all organisms.

16-2 Demand; 1; residual; 30.

16-3 Yellow green; 2.5.

16-4 Chlorine is a toxic substance.

16-5 Dose, mg/L = $\dfrac{400 \text{ lb/day}}{(5.55 \text{ MGD} \times 8.34)} = 8.64 \text{ mg/L}$

16-6 7.1 mg/L

16-7 May be required by plant's permit because chlorine is very toxic to aquatic organisms and must be removed to prevent stream damage.

16-8 HTH/day $\times$ 0.42 Available Chlorine = 147 lb chlorine/day

16-9 $$\frac{290 \text{ lb chlorine/day}}{0.69\% \text{ avail chlorine} \times 8.34 \text{ lb/gal} \times 1.18} = 42.7 \text{ gal/day}$$

16-10 $$\frac{45.7 \text{ lb/day} \times 1.10 \times 365 \text{ days}}{2,000 \text{ lb/container}} = (9.2) \, 10 \text{ containers}$$

16-11 Chlorine and its by-products (i.e., chloramines) are very toxic.

Chapter 17 Review Questions:

17-1 $30 \text{ min/cyc.} \times 24 \text{ h/3 h/cyc.} \times 65 \text{ gpm} \times 8.34 \text{ lb/gal} \times .052 \times .66 = 4,465 \text{ lb/day}$

17-2 Gravity thickeners, flotation thickeners, sludge concentrators.

17-3 Primary sludge.

17-4 Aerobic digesters, anaerobic digesters, composting, heat treatment, chlorine oxidation, lime stabilization.

17-5 Sand drying beds, vacuum filtration, belt filtering, centrifuges, incineration.

17-6 Drainage and evaporation.

17-7 $$\frac{41 \text{ mg} \times 52,000 \text{ gal} \times 3.785 \text{ L/gal}}{1 \text{ L} \times 454 \text{ g/lb} \times 1,000 \text{ mg/gram}} = 18 \text{ lb}$$

17-8 $$\frac{72,000 \text{ gal}}{2,780 \text{ gpd}} = 26 \text{ days}$$

17-9 90–95°F

17-10 $$\frac{335 \text{ mg/L}}{1,840 \text{ mg/L}} = 0.18 \text{ lb}$$

17-11 $$\frac{0.785 \times 45 \text{ ft} \times 45 \text{ ft} \times 22 \text{ ft} \times 7.48 \text{ gal/ft}^3}{4,800 \text{ gpd}} = 54.5 \text{ days}$$

17-12 $$\frac{(0.70 - 0.47) \times 100}{[0.70 - (0.70 \times 0.47)]} = 62\%$$

17-13 Area = $3.14 \times 10 \text{ ft} \times 8.4 \text{ ft} = 263.8 \text{ ft}^2$

$$\frac{30 \text{ gal}}{1 \text{ min}} \times \frac{60 \text{ min}}{1 \text{ h}} \times \frac{8.34 \text{ lb}}{1 \text{ gal}} \times \frac{12\%}{100\%} = 1801.4 \text{ lb/h}$$

$$\frac{1801.4}{263.8} = 6.8 \text{ lb/h/ft}^2$$

17-14 Returned to the plant for treatment.

17-15 Reduce sludge volume, stabilize organic matter, and recover organic matter for use in the plant.

Chapter 18 Review Questions:

18-1 15 minutes

18-2 $$\text{DO, mg/L} = \frac{(7.1 \text{ mL} - 0.0 \text{ mL}) \times 0.025 \text{ N} \times 8000}{200 \text{ mL}} = 7.1 \text{ mg/L}$$

18-3 A sample collected all at one time. Representative of the conditions only at the time taken.

18-4 10 mL/L of copper sulfate—sulfanic acid solution.

18-5 For pH, dissolved oxygen, total residual chlorine, fecal coliform, and any test required by NPDES permit for grab sample.

18-6 At different times to allow for the time it takes for wastewater to pass through treatment units.

18-7 A series of samples collected over a specified period of time in proportion to flow.

18-8 Composite samples reflect conditions in wastewater over a period of time.

18-9 Collect from well mixed locations; clearly mark sampling points; easy location to read; no large or unusual particles; no deposits, growths, or floating materials; corrosion-resistant containers; follow safety procedures; test samples as soon as possible.

18-10 A meter, reference electrode, and glass electrode.

18-11 USEPA Regulation (40 CFR 136) and the plant's permit.

18-12 USEPA

18-13 $\dfrac{3{,}000 \text{ mL}}{24 \text{ samples} \times 7.66 \text{ MGD}} = 16.3$

18-14 mL of sample $= 100 \times 4.66$ MGD $= 466$ or 470 mL

18-15 48 hours when preserved using refrigeration at 4°C.

18-16 day.

18-17 In the CBOD test, the nitrogenous oxygen demand is eliminated.

18-18 To ensure healthy organisms are available.

18-19 0.6 mg/L to 1.0 mg/L

18-20 Refrigerate at 4°C.

18-21 Absorption of water during cooling, contaminants, finger prints, etc.

Chapter 19 Review Questions:

19-1 14.3 mg/L $\times$ 5.89 MGD $\times$ 3.786 kg/MG/mg/L = 318.8 kg

19-2 Total 332.7
Average = 332.7/13 $\times$ 25.6 (no violation)

19-3 Week 1   370.9 kg/day
Week 2   273.6 kg/day
Week 3   1,062.4 kg/day (1 violation)
Week 4   499.8 kg/day

19-4

	BOD_5	TSS
Week 1	16.9 mg/L	15.0 mg/L
Week 2	13.8 mg/L	7.1 mg/L
Week 3	47.8 mg/L	47.3 mg/L
Week 4	24.0 mg/L	37.8 mg/L

19-5 No violation occurred.

19-6 Minimum = 4.5
Maximum = 7.2
25 exceptions (violations) occurred.

19-7 The licensed operator and the responsible official.

19-8 do.

APPENDIX B

Answers to Final Review Exam: Chapter 20

20-1 Prevent disease.
Protect aquatic organisms.
Protect water quality.

20-2 Dissolved and suspended.

20-3 Organic indicates matter that is made up mainly of carbon, hydrogen, and oxygen and will decompose into mainly carbon dioxide and water at 550°C.

20-4 Algae, bacteria, protozoa, rotifers, virus.

20-5 Carbon dioxide, water, more organisms, stable solids.

20-6 Toxic matter, inorganic dissolved solids, pathogenic organisms.

20-7 Raw effluent.

20-8 From body wastes of humans who have disease.

20-9 Disease-causing.

20-10 Domestic waste.

20-11 Industrial waste.

20-12 $3.14 \times 120 \text{ ft} = 377 \text{ ft}$

20-13 $180 \text{ ft} \times 110 \text{ ft} = 19,800 \text{ ft}^2$

20-14 $210 \text{ ft} \times 90 \text{ ft} \times 14 \text{ ft} = 264,600 \text{ ft}^3$

20-15 $[(2 \times 210 \text{ ft}) + (2 \times 90 \text{ ft})] \times 4 = 2,400 \text{ ft}$

20-16 $0.785 \times 120 \text{ ft} \times 120 \text{ ft} \times 18 \text{ ft} = 203,472 \text{ ft}^3$
$\phantom{0.785 \times 120 \text{ ft} \times 120 \text{ ft} \times 18 \text{ ft} =\ } 30,182.4 \text{ ft}^3$
$\phantom{0.785 \times 120 \text{ ft} \times 120 \text{ ft} \times 18 \text{ ft} =\ } 233,654.4 \text{ ft}^3$

20-17 $\dfrac{22 \text{ g} \times 100}{500 \text{ g}} = 4.4\%$

20-18 2.3 ft

20-19 $5,250 \text{ gal} \times 8.34 \text{ lb/gal} = 43,785 \text{ lb}$

20-20 $1,920 \text{ ft}^3 \times 7.48 \text{ gal/ft}^3 = 14,362 \text{ gal}$

20-21 $30 \text{ mg/L} \times 3.40 \text{ MGD} \times 8.34 \text{ lb/MG/mg/L} = 850.7 \text{ lb/day}$

20-22 $5 \text{ mg/L} \times 7.25 \text{ MGD} \times 3.785 \text{ kg/MG/mg/L} = 686 \text{ kg/day}$

20-23 $\dfrac{3,280 \text{ lb/day}}{3,250 \text{ mg/L} \times 8.34 \text{ lb/MG/mg/L}} = 0.121 \text{ MGD}$

20-24 $\dfrac{240 \text{ mg/L} \times 0.72 \text{ MGD} \times 8.34 \text{ lb/MG/mg/L}}{0.17 \text{ lb BOD}_5/\text{person/day}} = 8,477 \text{ people}$

20-25 $8.34 \text{ lb/gal} \times 1.1288 = 9.41 \text{ lb/gal}$

20-26 $\dfrac{(230 \text{ mg/L} - 22 \text{ mg/L}) \times 100}{230 \text{ mg/L}} = 90.4\%$

20-27 $\dfrac{(370 \text{ mg/L} - 18 \text{ mg/L}) \times 100}{370 \text{ mg/L}} = 95.1\%$

20-28 $\dfrac{(230 \text{ mg/L} - 170 \text{ mg/L}) \times 100}{230 \text{ mg/L}} = 26\%$

20-29 $\dfrac{(0.665 - 0.489) \times 100}{[0.665 - (0.665 \times 0.489)]} = 52\%$

20-30 $\dfrac{140' \times 100' \times 12' \times 2 \text{ tanks} \times 7.48 \text{ gal/ft}^3 \times 24 \text{ h/day}}{8.40 \text{ MGD} \times 1,000,000 \text{ MGD}} = 7.2 \text{ h}$

20-31 $\dfrac{40' \times 3' \times 2.5' \times 2 \text{ channels} \times 7.48 \text{ gal/ft}^3 \times 1,440 \text{ min/day}}{8.40 \text{ MGD} \times 1,000,000 \text{ gal/MG}} = 0.77 \text{ min}$

20-32 $\dfrac{115,000 \times 7.48 \text{ gal/ft}^3}{19,000 \text{ gpd}} = 45 \text{ days}$

20-33 To remove large objects.

20-34 Manual and mechanical cleaners.

20-35 Burial in an approved landfill; incineration.

20-36 Cutter must be sharpened and/or replaced when needed.
Cutter alignment must be adjusted as needed.

20-37 Because the cutters have been replaced and the alignment has been checked, the most likely cause is excessive solids in the plant effluent. Corrective actions would include identifying the source, implementing/creating a sewer use ordinance, and/or installing a bar screen upstream of the comminutor to decrease load it receives.

20-38 Grit is heavy inorganic matter. Sand, gravel, metal filings, egg shells, coffee grounds, etc.

20-39 $\dfrac{8.0 \text{ MGD} \times 1.55 \text{ cfs/MGD}}{3 \text{ channels} \times 2' \times 3'} = 0.7 \text{ fps}$

20-40 There is a large amount of organic matter in the grit. The aeration rate must be increased to prevent settling of the organic solids.

20-41 To remove settleable and flotable solids.

20-42 To remove the settleable solids formed by the biological activity.

20-43 $\dfrac{0.785 \times 90' \times 90' \times 12 \times 7.48 \text{ gal/ft}^3 \times 24 \text{ h/day}}{2.25 \text{ MGD} \times 1,000,000 \text{ gal/MG}} = 6.1 \text{ h}$

$\dfrac{2.25 \text{ MGD} \times 1,000,000 \text{ gal/MG}}{0.785 \times 90' \times 90'} = 354 \text{ gpd/ft}^2$

$\dfrac{2.25 \text{ MGD} \times 1,000,000 \text{ gal/MG}}{3.14 \times 90'} = 7,962 \text{ gpd/ft}$

20-44 $\dfrac{1,500' \times 1,100'}{43,560 \text{ ft}^3/\text{acre-feet}} = 37.9 \text{ acres}$

20-45 $\dfrac{1,500' \times 1,100' \times 4.1'}{43,560 \text{ ft}^3/\text{acre-feet}} = 155.3 \text{ acre-feet}$

20-46 $0.44 \text{ MGD} \times 3.069 \text{ acre-ft/MG} = 1.35 \text{ acre-feet/day}$

20-47 $0.44 \text{ MGD} \times 36.8 \text{ acre-in./MG} = 16.2 \text{ ac-in./day}$

20-48 $\dfrac{155.3 \text{ acre-feet}}{1.35 \text{ acre-ft/day}} = 115 \text{ day}$

20-49 $\dfrac{16.2 \text{ ac-in./day}}{37.9 \text{ acres}} = 0.43 \text{ in./day}$

20-50 $\dfrac{380 \text{ mg/L} \times 0.44 \text{ MGD} \times 8.34 \text{ lb/MG/mg/L}}{37.9 \text{ acres}} = 36.8 \text{ lb BOD}_5/\text{acre/day}$

20-51 $\dfrac{970 \text{ mg/L} \times 0.039 \text{ MGD} \times 8.34 \text{ lb/MG/mg/L}}{0.17 \text{ lb BOD}_5/\text{P.E.}} = 1{,}855 \text{ P.E.}$

20-52 Stabilization pond, oxidation pond, polishing pond.

20-53 Settling, anaerobic digestion of settled solids, aerobic/anaerobic decomposition of dissolved and colloidal organic solids by bacteria producing stable solids and carbon dioxide, photosynthesis production of oxygen by algae.

20-54 Summer effluent is high in solids (algae) and low in BOD$_5$.
Winter effluent is low in solids and high in BOD$_5$.

20-55 Eliminates wide diurnal and seasonal variation in pond DO.

20-56 Increases during the daylight hours and decreases during darkness.

20-57 Reduces fecal coliform, BOD$_5$, TSS, and nutrient levels.

20-58 $\dfrac{2.40 \text{ MGD} \times (1.0 + 1.5)}{2} = 3.0 \text{ MGD}$

20-59 $\dfrac{2.40 \text{ MGD} \times (1.0 + 0.70)}{2} = 2.04 \text{ MGD}$

20-60 $\dfrac{2.40 \text{ MGD} \times (1.0 + 1.5) \times 1{,}000{,}000 \text{ gal/MG}}{0.785 \times 90' \times 90' \times 2} = 472 \text{ gpd/ft}^2$

20-61 $\dfrac{2.40 \text{ MGD} \times (1.0 + 0.75) \times 1{,}000{,}000 \text{ gal/MG}}{0.785 \times 90' \times 90' \times 2} = 330 \text{ gpd/ft}^2$

20-62 $\dfrac{196 \text{ mg/L} \times 240 \text{ MGD} \times 8.34 \text{ lb/MG/mg/L} \times 1{,}000}{2 \text{ filters} \times 0.785 \times 90' \times 90' \times 8'} = 38.6 \text{ lb/BOD}_5/1{,}000 \text{ ft}^3$

20-63 Distribution system—to distribute the hydraulic and organic loading evenly over the filter media.
Media—to support the biological growth.
Underdrains—to collect and remove treated wastewater and sloughings from the filter.
To provide ventilation.

20-64 Standard (best effluent quality)
High rate
Roughing

20-65 $250{,}000 \text{ ft}^2 + 220{,}000 \text{ ft}^2 + (6 \times 150{,}000 \text{ ft}^2) = 1{,}370{,}000 \text{ ft}^2$

$\dfrac{840 \text{ MGD} \times 1{,}000{,}000 \text{ gal/MG}}{1{,}370{,}000 \text{ ft}^2} = 6.1 \text{ gpd/ft}^2$

20-66 $\dfrac{190 \text{ mg/L} \times 8.40 \text{ MGD} \times 8.34 \text{ lb/MG/mg/L} \times 1{,}000}{1{,}370{,}000 \text{ ft}^2} = 9.72 \text{ lb BOD}_5/\text{day}/1{,}000 \text{ ft}^2$

20-67 $190 \text{ mg/L} - (150 \text{ mg/L} \times 0.55) = 107 \text{ mg/L}$

$\text{SOL} = \dfrac{107 \text{ mg/L} \times 8.40 \text{ MGD} \times 8.34 \text{ lb/MG/mg/L}}{1{,}370{,}000 \text{ ft}^2} = 5.5 \text{ lb BOD}_5/\text{day}/1{,}000 \text{ ft}^2$

20-68 Disks covered with biological growth rotate in wastewater. Organisms collect food during submergence. Oxygen is transferred during exposure to air. Organisms oxidize organic matter. Waste products and sloughings are discharged to wastewater flow for removal in settling tank.

20-69 The use of fixed film biological organisms.

20-70 Normal—gray, shaggy.
High sulfur—chalky, white.

20-71 $\dfrac{1{,}750\ \text{mL} \times 100}{2{,}000\ \text{mL}} = 88\%$

20-72 $\dfrac{1{,}050\ \text{mL}}{2\ \text{L}} = 525\ \text{mL/L}$

20-73 $\dfrac{1{,}750\ \text{mL}}{2\ \text{L}} = 875\ \text{mL/L}$

$\dfrac{875\ \text{mL} \times 1{,}000}{2{,}750\ \text{mg/L}} = 318$

20-74 $\dfrac{175\ \text{mg/L} \times 2.10\ \text{MGD} \times 8.34}{1{,}850\ \text{mg/L} \times 1.255\ \text{MG} \times 8.34} = 0.16\ \text{BOD}_5 / \text{lb MLVSS}$

20-75 $\dfrac{2{,}750\ \text{mg/L} \times (1.255\ \text{MG} + 0.88\ \text{MG}) \times 8.34}{(6{,}120\ \text{mg/L} \times 0.090\ \text{MGD} \times 8.34) + (18\ \text{mg/L} \times 2.10\ \text{MGD} \times 8.34)} = 10.0\ \text{days}$

20-76 Increase waste rate.

20-77 Decrease; decrease; decrease; increases; increase.

20-78 320 lb/hypochlorite × 0.45 available chlorine = 144 lb/chl/day

20-79 $\dfrac{285\ \text{lb chlorine/day}}{0.679\%\ \text{Available Chlorine} \times 8.34\ \text{lb/gal} \times 1.18} = 42.7\ \text{gal/day}$

20-80 $\dfrac{45.8\ \text{lb/day} \times 1.10 \times 365\ \text{days}}{2{,}000\ \text{lb/containers}} = (9.2)\ 10\ \text{containers}$

20-81 $\dfrac{6\ \text{containers} \times 2{,}000\ \text{lb/container}}{6.2\ \text{mg/L} \times 2.25\ \text{MGD} \times 8.34\ \text{lb/MG/mg/L}} = 88\ \text{days}$

20-82 $\dfrac{38\ \text{lb/day} \times 1.11 \times 365\ \text{days}}{150\ \text{lb/cylinder}} = 102.6\ (103\ \text{cylinders})$

103 cylinders × 150 lb/cyl × $0.17/lb × 1.075 = $2,823.49

20-83 Disinfection destroys pathogenic organisms.
Sterilization destroys all organisms.

20-84 demand; 1; residual; 30.

20-85 yellow green; 2.5.

20-86 30 min/cyc × 24 h/3 h/cyc × 70 gpm × 8.34 lb/gal × 0.051 × 0.66 = 4,716 lb/day

20-87 $\dfrac{41\ \text{mg} \times 63{,}000\ \text{gal} \times 3.785\ \text{L/gal}}{1\ \text{L} \times 454\ \text{g/lb} \times 1{,}000\ \text{mg/gram}} = 21.5\ \text{lb}$

20-88 $\dfrac{73{,}500\ \text{gal}}{2{,}750\ \text{gpd}} = 27\ \text{days}$

20-89 90–95°F
Maximum change should be one (1) degree per day.

20-90 $\dfrac{340\ \text{mg/L}}{1{,}830\ \text{mg/L}} = 0.19$

Answers to Final Review Exam: Chapter 20

20-91 $\dfrac{0.785 \times 50' \times 50' \times 25' \times 7.48 \text{ gal}/\text{ft}^3}{6{,}000 \text{ gpd}} = 61.2 \text{ days}$

20-92 $\dfrac{(0.72 - 0.48) \times 100}{[0.72 - (0.72 \times 0.48)]} = 64.1\%$

20-93 The licensed operator and the responsible official.

20-94 National Pollutant Discharge Elimination System.

20-95 By increasing the primary sludge pumping rate or by adding dilution water.

20-96 Flow rate of the sludge is going into the unit.
Pounds or kg of solids in the influent.

20-97 Thermal conditioning.

20-98 7.0 pH.

20-99 Either because the microorganisms have been killed or are absent.

20-100 The time to do the test, 3 hours versus 5 days.

20-101 1 foot per second.

20-102 Grit chambers.

20-103 Density of sludge.

20-104 Anaerobic conditions within the filter.

20-105 Decrease in alkalinity.

APPENDIX C

Commonly Used Formulae in Wastewater Treatment

Parameter	Formula
Area, ft^2	
Rectangle	Width, ft × Length, ft [(W)(L)]
Circle	.785 (Diameter, ft)^2 [(.785)D^2]
Volume, ft^3	
Rectangle	Width, ft × Length, ft × Height [(W)(L)(H)]
Cylinder	.785 (Diameter, ft)^2 (Height, ft) [(.785)D^2H]
Cone	(1/3)(π)(Radius, ft)^2 (Height, ft) [.33 R^2H]
Sphere	(4/3)(π)(Radius, ft)^3 [4.16 R^3]
Flow, cfs	(Velocity, ft/sec)(Surface Area, ft^2) [(V)(A)]
Pounds	(Flow, MGD)(con, mg/L)(8.34 lb/gal) [(Q)(mg/L)(8.34)]
SVI, mL/g	$\dfrac{\text{Volume}}{\text{MLSS concentration}} \times 100$
Circumference, ft	(π)(Diameter)
Detention Time, h	$\dfrac{(\text{Volume, gal})(24\,\text{h/day})}{\text{Flow, gpd}}$, [V/Q]
Surface Loading Rate, gpd/ft^2	$\dfrac{\text{Flow, gpd}}{\text{Area, ft}^2}$, [Q/A]
Organic Loading Rate	$\dfrac{\text{lb of BOD}}{\text{lb MLVSS}}$
Sludge Age	$\dfrac{\text{MLSS in Aeration Tank}}{\text{SS in Primary Effluent}}$
MCRT	$\dfrac{\text{SS in Secondary System}}{\text{WAS/day + SS in effluent/day}}$
Volatile Solids Reduction	$\dfrac{\text{In} - \text{Out}}{\text{In} - (\text{In} \times \text{Out})} \times 100$
Chlorine Dose	Chlorine Demand + Chlorine Residual
Removal, %	In − Out × 100

Index

activated sludge, 5, 151–166
advanced wastewater treatment, 5
aerated pond, 134
aerated systems, 115–116
aerobic, 5
aerobic digestion, 189–192
aerobic ponds, 134
aerobic processes, 77
algae, 75–76
alkalinity, 71
amperometric direct titration, 212–213
amoeba, 76
anaerobic, 5
anaerobic digestion, 192–199
anaerobic ponds, 134
anaerobic processes, 77–78
animal waste, 70
anoxic, 6
anoxic processes, 78–79
area, 37–39, 91
arithmetic average, 14–15
available chlorine, 175–176
average daily limitation, 229
average hourly limitation, 229
average monthly discharge limitation, 6
average monthly limitation, 229
average weekly discharge limitation, 6
average weekly limitation, 229

bacteria, 76
bar screen, 109–110
barminution, 112
base number, 16
biochemical cycles, 81–82
biochemical oxygen demand (BOD), 6–7
biological processes, 77–79
biosolids, 6
BOD$_5$ calculation, 221–223
brake horsepower, 98
buffer, 6

calculator, 10
carbon cycle, 82
carbonaceous biochemical oxygen demand, 6
centrifugal pumps, 94

centrifuges, 202
chemical addition, 119
chemical oxygen demand (COD), 6, 72
chlorine, 169
chlorine demand, 174
chlorine dose, 175
chlorine feed rate, 174–175
chlorine ordering, 177–178
chlorination, 170–181
circumference, 34–37
clarifier, 6, 123–125
coliform, 6
color, 71
combined sewer, 6
combined wastewater, 70
comminution, 6, 112
common factor, 10
composite sample, 6, 207–208
conditional equation, 27
contact time, 169
conversion factors, 51–59
conversion table, 53
conventional seals, 95
cross-sectional area, 37
cyclone degritter, 116

daily discharge, 6, 229
daily maximum discharge, 6
dechlorination, 181
decimal, 12–13
delayed inflow, 7
demand, 169
density, 86–87
detention time, 6
dewatering, 6, 199–202
dimensional analysis, 22–26
direct flow, 7
discharge monitoring report (DMR), 6, 230
discharge permit, 230
disinfection, 169–181
dissolved gases, 72
dissolved oxygen (DO), 6, 216–219
dissolved solids, 72
diurnal, 71
dividend, 10

domestic wastewater, 70
    characteristics of, 72–73
dose, 169
DPD-FAS titration, 212
DPD spectrophotometric, 210–212
dry well, 103

effluent, 6
effluent limitation, 6, 230
electrical power, 98
elemental chlorine, 171
equations, 27–31
    axioms, 28
    defined, 27
estimated return rate, 162
exponent, 16

factor, 10
facultative, 6
facultative ponds, 134
fecal coliform, 7
feed rate, 169
filter presses, 200–201
fill and draw, 116–117
fixed film systems, 131–132
flagellates, 76
floc, 7
flotation thickening, 188–189
flow, 71, 91
flow equalization, 119
flow measurement, 116–118
flume, 7, 118
food-to-microorganism ratio, 164–165
force, 88
force main collection system, 101
fractions, 12–13
free-swimming ciliates, 76
friction head, 85, 89–90

geometric mean, 42–43
grab sample, 7, 206
gravity collection system, 101
gravity thickening, 188
grit, 7
grit removal, 113–115
groundwater runoff, 8
growth curve, 79–80

head, 85, 89–90
horsepower, 97
household waste, 70
human waste, 70
hydraulics, 85–91
hydraulic detention time, 65–66
hypochlorite, 170–171

Imhoff tank, 123
incinerators, 202
infiltration, 7

infiltration/inflow, 7
influent, 7
industrial waste, 70
industrial wastewater, 7
inorganic, 7
inorganic content, 72
integer, 9
iodometric direct titration method, 213–215

kinetic pumps, 93

law of conservation of mass, 44
license, 7
logarithm method, 42

manually cleaned screens, 110–111
mass balance, 44–47
    for settling tanks, 45–46
    using BOD removal, 46
maximum daily discharge, 229
maximum discharge, 230
mean cell residence time (MCRT), 7, 165–166
mechanical seals, 95
mechanically cleaned screens, 111
milligrams/liter (mg/L), 7
mixed liquor, 7
mixed liquor suspended solids (MLSS), 7
mixed liquor volatile suspended solids (MLVSS), 7
moving average, 43–44
modified Winkler method, 219–220
motor horsepower, 98

nitrogen compounds, 72
nitrogenous oxygen demand (NOD), 7
NPDES permit, 7, 230–234
nutrients, 7
nutrient cycle, 82–83

odor, 71
organic, 7
organic solids, 72
oxidation pond, 134

part per million (ppm), 8
pathogenic, 8
percent, 13–14
percent total solids (%TS), 127
percent volatile matter reduction, 62
performance efficiency, 61–62
perimeter, 35
pH, 72, 209–210
phosphorus, 72
photosynthesis, 79
point source, 8, 230
polishing pond, 134
positive displacement pump, 93–94
powers of ten, 15–18
preaeration, 118
preliminary treatment, 109–119

Index

pressure, 85, 88–91
prime number, 10
process residuals, 183–202
product, 10
proportion, 31–33
protozoa, 76
pumping station, 103–105
pump calculation, 96–99
pumps, 93–99

quotient, 10

ratio, 31
return activated sludge solids (RASS), 8
rotating biological contactors, 143–146
rotifers, 76
rounding a number, 26–27

sample preservation, 208
sampling and testing, 205–226
sand drying beds, 199
sanitary wastewater, 8, 70
scientific notation, 19–21
screening, 109–112
scum, 8
secondary treatment, 131–132
sedimentation, 121–129
self-purification, 80–81
septic, 8
sequence of operations, 11–12
settleability, 8
settled sludge volume, 8, 156–157, 161–162
sewage, 8
shaft lubrication, 95
shredding, 112
sludge, 8
sludge pumping, 126
sludge pumping calculation, 183–187
sludge retention time (SRT), 8
sludge thickening, 187–189
sludge volume index (SVI), 8, 163
solids, 72
solid concentration, 166
solids sampling and testing, 223–226
specific gravity, 87–88
stabilization, 189–199

stalked ciliated protozoa, 76
static head, 85, 89
steady inflow, 7
storm sewer, 8
stormwater, 8, 70
stormwater runoff, 70
specific gravity, 59
sulfur cycle, 83
supernatant, 8
surface area, 37
suspended solids, 72
suspended growth systems, 132

temperature, 71
temperature conversions, 51–53
total dynamic head, 85, 90
total inflow, 7
total residual chlorine, 210
treatment ponds, 132–136
trickling filters, 137–142

unit of measurement, 33–34
unknown, 27

vacuum filters, 199–200
vacuum system, 101–103
velocity, 85, 91
velocity head, 85, 90
virus, 77
volume, 39–41

waste activated sludge, 163–164
waste activated sludge solids (WASS), 8
wastewater, 8
wastewater characteristics, 70–73
    physical characteristics, 71
    chemical characteristics, 71
wastewater collection systems, 101–107
wastewater testing, 208–226
water horsepower, 97
weirs, 117–118
weir loading rate, 125–126
wet well, 103–104
wet well calculations, 105–107

zoogleal slime, 8